T0188883

Studies in Russia and East Europe

This series includes books on general, political, historical, economic and cultural themes relating to Russia and East Europe written or edited by members of the School of Slavonic and East European Studies, University College London, or by authors working in association with the School.

Titles include:

Roger Bartlett and Karen Schönwälder (*editors*)
THE GERMAN LANDS AND EASTERN EUROPE

John Channon (*editor*)
POLITICS, SOCIETY AND STALINISM IN THE USSR

Stanislaw Eile
LITERATURE AND NATIONALISM IN PARTITIONED POLAND, 1795–1918

Jane Grayson, Arnold Macmillan and Priscilla Meyer (*editors*)
NABOKOV'S WORLD
Volume 1: The Shape of Nabokov's World
Volume 2: Reading Nabokov

Celia Hawkesworth (*editor*)
A HISTORY OF CENTRAL EUROPEAN WOMEN'S WRITING

Celia Hawkesworth, Muriel Heppell and Harry Norris (*editors*)
RELIGIOUS QUEST AND NATIONAL IDENTITY IN THE BALKANS

Rebecca Haynes
ROMANIAN POLICY TOWARDS GERMANY, 1936–40

Geoffrey Hosking and Robert Service (*editors*)
RUSSIAN NATIONALISM, PAST AND PRESENT

Lindsey Hughes (*editor*)
PETER THE GREAT AND THE WEST
New Perspectives

Krystyna Iglicka and Keith Sword (*editors*)
THE CHALLENGE OF EAST–WEST MIGRATION FOR POLAND

Andres Kasekamp
THE RADICAL RIGHT IN INTERWAR ESTONIA

Stephen Lovell
THE RUSSIAN READING REVOLUTION

Marja Nissinen
LATVIA'S TRANSITION TO A MARKET ECONOMY

Danuta Paszyn
THE SOVIET ATTITUDE TO POLITICAL AND SOCIAL CHANGE IN CENTRAL AMERICA, 1979–90

Vesna Popovski
NATIONAL MINORITIES AND CITIZENSHIP RIGHTS IN LITHUANIA, 1988–93

Alan Smith
THE RETURN TO EUROPE
The Reintegration of Eastern Europe into the European Economy

Jeremy Smith
THE BOLSHEVIKS AND THE NATIONAL QUESTION, 1917–23

Jeanne Sutherland
SCHOOLING IN THE NEW RUSSIA

Kieran Williams and Dennis Deletant
SECURITY INTELLIGENCE SERVICES IN NEW DEMOCRACIES
The Czech Republic, Slovakia and Romania

Studies in Russia and East Europe
Series Standing Order ISBN 978-0-333-71018-0
(*outside North America only*)

You can receive future titles in this series as they are published by placing a standing order. Please contact your bookseller or, in case of difficulty, write to us at the address below with your name and address, the title of the series and the ISBN quoted above.

Customer Services Department, Macmillan Distribution Ltd, Houndmills, Basingstoke, Hampshire RG21 6XS, England

Religious Quest and National Identity in the Balkans

Edited by

Celia Hawkesworth
Senior Lecturer in Serbian and Croatian Studies
University College London

Muriel Heppell
Emeritus Reader in the Medieval History of Orthodox Eastern Europe
University of London

and

Harry Norris
Emeritus Professor of Arabic and Islamic Studies
University of London

 in association with
School of Slavonic and East European Studies
University College London

First published 2001 by
PALGRAVE
Houndmills, Basingstoke, Hampshire RG21 6XS and
175 Fifth Avenue, New York, N. Y. 10010
Companies and representatives throughout the world

PALGRAVE is the new global academic imprint of
St. Martin's Press LLC Scholarly and Reference Division and
Palgrave Publishers Ltd (formerly Macmillan Press Ltd).

ISBN 978-1-349-41772-8   ISBN 978-0-230-52333-3 (eBook)
DOI 10.1057/9780230523333

This book is printed on paper suitable for recycling and made from fully
managed and sustained forest sources. Logging, pulping and manufacturing
processes are expected to conform to the environmental regulations of the
country of origin.

A catalogue record for this book is available
from the British Library.

Library of Congress Cataloging-in-Publication Data
Religious quest and national identity in the Balkans / edited by
Celia Hawkesworth, Muriel Heppell, Harry Norris.
        p. cm. — (Studies in Russia and East Europe)
    Includes bibliographical references and index.

    1. Balkan Peninsula—Religion. 2. Balkan Peninsula—Civilization.
3. Balkan Peninsula—Ethnic relations—History. 4. Ethnic
relations—Religious aspects—History. 5. Nationalism—Balkan
Peninsula—History. I. Hawkesworth, Celia, 1942– II. Heppell, Muriel.
III. Norris, H. T. IV. Series.

BL980.B28 R45 2001
200'.9496—dc21
                                                        2001021738

10   9   8   7   6   5   4   3   2   1
10   09 08 07 06 05 04 03 02 01

For Sonia I. Kanikova

Contents

List of Illustrations

Foreword

This book grew out of a series of seminars held at SSEES between January and May 1997, followed by a conference, 30 June–2 July 1997. The original idea was conceived at a meeting between Celia Hawkesworth (SSEES), Judith Herrin (King's College, London) and Sonia Kanikova (SSEES), in the summer of 1996. The organising committee was later joined by Florentina Badalanova (SSEES), Ger Duijzings (SSEES), Harry Norris (SOAS) and Yuri Stoyanov (Warburg Institute). But the project was put into practice very largely thanks to the dedication and commitment of Sonia Kanikova, who had laid the groundwork and completed the bulk of the organisational work before her resignation from SSEES in January 1997. The aim of the project was to cover as much as possible of the main aspects of religion and religious life in the Balkans, from pre-Christian traditions, monasticism, mysticism, the interaction between Eastern Orthodoxy and Roman Catholicism, the diffusion of Islam, the relationship between Islam and Christianity, Judaism, the role of religion in shaping national identities in the Balkans and its place in Balkan society today. Many of these issues were discussed in the seminars and conference, which also provided a forum for wide-ranging and fruitful discussion. Inevitably a volume of this kind cannot hope to offer both comprehensive coverage and detailed analysis of such a broad and complex area, and many fascinating topics – such as Catholicism and Judaism – are barely touched on, not for lack of will, but simply as a result of the vagaries of conference attendance, sudden illness, and which particular papers were received by the editors. We hope, nevertheless, that we have achieved a reasonable balance and that the individual chapters and general bibliography will encourage readers to explore individual themes further for themselves.

The main focus of the volume, as it has gradually taken shape, is on aspects of the two main Balkan religions, Christianity and Islam. As Harry Norris stresses in his introduction, and as this volume illustrates, Balkan religion has been described by many researchers – historians, archaeologists, anthropologists – as a classic example of religion being shaped by heterodoxy and popular superstitions. The variations of interaction between the two great monotheistic religions described in these pages are underlain by survivals of earlier beliefs, which permeate the particular forms of religious life as they have evolved in the Balkans.

The volume endeavours to give a sense of those survivals as well as specific features of the main religious movements. In order to emphasise this aim, the chapters are arranged broadly chronologically: the first chapters concern pre-Christian elements, archaeological survivals, popular religion and the complex mixture of ethnic and religious groups before the Ottoman invasion. Later sections are devoted to issues in Orthodoxy and Islam, while the last two chapters bring us up to the present, with a discussion of the involvement of religion in politics and the manipulation of belief for political ends.

Focusing, then, on Christianity and Islam as they were manifested in the Balkans, the Introduction by Heppell and Norris deals with the history and the present reality of these two dominant religions in the region. Heppell gives an account of one of the first specific strands of religious life in the Balkans: the Bogomil heresy, and the long-held belief that there was a close connection between it and the emergence, in the thirteenth century, of the 'Bosnian Church'. While there are conflicting views of the Bosnian Church, Heppell suggests that they are not necessarily incompatible. She stresses that, despite the Schism of 1054, until the Ottoman conquest, there was considerable contact between members of both the Eastern and the Western churches. This is one of many examples of the ambiguity and accommodating nature of religious belief in the Balkans, so vividly illustrated in Noel Malcolm's chapter (Chapter 6) on syncretism. In keeping with the fundamental character of the region as a borderland, the edges of many aspects of religious life are blurred. Heppell also gives an account of the 'devshirme' or 'tribute in blood', a feature of Ottoman rule mentioned in several chapters. She ends by emphasising the fact that recent research shows that converts to Islam in the Balkans came from all religious groups.

This is borne out by Harry Norris in the second section of the Introduction, in which he describes the particular nature of Islam in the Balkans as being 'shaped by a thick underlay' of Christian and pre-Christian belief. He stresses that for Balkan Muslims, religion is a way of life, an attitude of mind and mark of identity, rather than a creed or dogma. In turn, he underlines the tendency of Balkan Islam to draw on the spiritual power of the faiths which preceded it. He cites the example of the conviction of Bosnian Muslims in particular that their faith is an organic growth from past religious beliefs rather than an alien transplant. Norris discusses the academic achievements of Islam in Bosnia, before comparing it with Islam in the Albanian-speaking communities, where there are many survivals of pagan (Illyrian?) beliefs. Balkan Islam tends to be viewed by Muslims in the wider world, who have drawn their

faith from Arab trends, as somewhat apart: the intellectual and speculative inclinations of Balkan Islam have been nourished, through the
Turks, by Iranian sources. The legacy of Sufism is particularly strong, as
demonstrated later in the volume by John Norton's discussion of the
prominence of the Bektashi order and by the respect for such traditions
displayed even by a Marxist such as Enver Hoxha. As Norris stresses in
his concluding remarks: 'Islamic "heterodoxy" displays a leavening and
a dynamic impulse in the Islam of the Balkans, both at a popular and at
an intellectual and spiritual level. The variety, the influences of a cultural kind which it brought with it from the Orient and disseminated
deep within South-Eastern Europe cannot be assessed by a simple study
of local communities, however profound that study.' This volume seeks
to illustrate something of the variety of influences of which not only
Islam, but the religious life of the Balkans altogether is composed.

In his 'Prolegomenon', John Nandriş takes the long reflective look of
an archaeologist back at the layers of belief that make up the religious
heritage of the Balkans. He suggests that much contemporary behaviour
and what he calls 'the pseudo-religion of Communism' can be properly
explained only through awareness of these layers. He stresses that Balkan religion and culture are to some degree 'mythological codifications
of experience' and that all the major religions sought to incorporate into
their practice those elements of earlier beliefs which could not be
entirely eradicated. Nandriş usefully reminds us also of the linguistic
ambiguity of the Greek term 'ethnos', and raises the crucial issue of
religion being often used in the Balkan context to explain ethnic identity and conflict – an issue which Mitja Velikonja discusses in the concluding chapter of the volume. Nandriş describes this as an essentialist
view, which has often been exploited by politicians to obscure their own
unsubtle ends. His clear formulation may serve as a motto for this whole
volume: 'Balkan religions, like the cultures of their practitioners, often
exhibited an ability to reconcile disparate elements harmoniously
whether through syncretism or symbiosis. It is the politicians and the
purists seeking exclusive solutions who have so often created strife.'

Florentina Badalanova examines one instance of the interaction of
popular belief and the 'religions of the book', stressing that an understanding of their correlation is important for a coherent picture of the
cultural and social processes at work in South-East Europe today. She
examines the convergence of Christian and Muslim elements at the
level of folk religion, pointing out that Judaic influences may also
be traced. She uses the example of the story of the sacrifice of Abraham
as told in the Bulgarian folk tradition to demonstrate not only ways in

which the three traditions are reconciled at this level, but also their common origins. Badalanova shows the way in which the Slavonic and Balkan oral tradition still 'remembers' prototexts that generated not only the Bible and the Koran, but also some midrashic texts.

Elissaveta Moussakova considers an aspect of the Bulgarian transition from paganism to Christianity as it is illustrated by some typical examples of the link between art and doctrine. She suggests that conversion stemmed from religious pluralism tending to monotheistic doctrines, probably through contacts with Byzantine culture, but, in the case of Bulgaria, also proto-Bulgarian culture itself. This chapter offers a specific illustration of a phenomenon seen throughout the Balkans: notably the existence of the remains of places of pagan worship, later converted into Christian temples which suggest that a certain organisation of religious life and practice preceded conversion. Moussakova explores the relationship between Christianity and paganism, official doctrine and 'vernacular' Christianity by looking at some monuments from pagan times that continued to function after the conversions: stone pillars with inscriptions, a sacred rock relief and the cross as symbolising the defeat of paganism. She also considers graffiti of the cross itself as examples of the transposition of the central Christian symbol to the level of folklore, where it was adapted to a meaningful role in the community in the light of pagan traditions.

The theme of the interplay between the official churches and popular belief and the ambiguity of some manifestations of belief is continued in the next two chapters. Bernard Hamilton's discussion of early manifestations of dualist heresies in Constantinople focuses particularly on the Bogomils, who played such an important part in the religious life of the central Balkans and especially Bosnia. Adherents of Bogomilism were of great concern to the Byzantine authorities because their initiates were hard to distinguish from Orthodox monks and they were therefore well placed to infiltrate the official church. Hamilton also examines relations between these groups and the Cathars in southern Europe. The next chapter, by Ioan-Aurel Pop, describes the Hungarian kingdom in the late thirteenth and fourteenth centuries, where there was evidence of both the Western and the Eastern traditions, alongside Judaism, Islam and several heresies – while the general population was largely pagan. This account may easily be applied to other parts of the Balkans at this period. Pop considers the policies of the Catholic Church on the territory of Hungary towards 'schismatics' and non-Christians, aimed at creating Catholic unity of the entire people. He gives a detailed account of the mechanisms of establishing homogeneity. In the Balkan context, it is possible to interpret such a procedure as an early instance of the impulse to reinforce a power base through purification and 'ethnic cleansing' that has been such a shameful feature of the history of the central Balkans in the late twentieth century. As so many of the contributions to this volume demonstrate, the natural condition of the region is an inextricable mix of peoples, traditions, cultures and belief. Pop's list of the mixed population found by the Hungarians on their invasion of Pannonia offers a snapshot which is typical of the whole region: there were Slavs, Slavicised Bulgarians, Romanians, Germanic peoples, Byzantines, old groups of Avars, Khazars (maybe Szeklers) who had arrived in advance of the Hungarians, 'Latins'; there were peoples conquered by the Hungarians: Slovaks, Serbs, new Bulgarians, Ruthenians; groups which had arrived through migrations and colonisation from the West: urban Germans, Flemings, Saxons, and from the East: North-Iranians, Khorezmians, Caucasian Alanians (Sarmatians), Bashkirs, Petchenegs, Udae, Cumanians, Jews, Iasians; and people who had come through marriage alliances, dynastic unions or diplomacy, such as Croatians, Bosnians, and Italians from Dalmatia. As this list of peoples living in one small region of the Balkans suggests, any effort to impose homogeneity or to disentangle the ethnic origins of its inhabitants in the name of 'ethnic purity' is a barren and nonsensical project.

After these chapters considering the early stages of the merging of traditional pagan belief with Christianity, Noel Malcolm analyses the widespread phenomenon of 'crypto-Christianity' and religious 'amphibianism' in the Balkans under Ottoman rule. These terms are used to describe small communities scattered through the Ottoman lands which adhered outwardly to Islam, while privately retaining vestiges of a Christian, Catholic, faith. Malcolm provides also a useful overview of the conditions of religious life in Ottoman-ruled Christian or former Christian territories. He warns against a simplistic account of unproblematic interaction between Christianity and Islam, which has lately tended to replace the earlier crude religious-nationalist versions, promoting a view of oppressed national churches and oppressive Islam. With precision and clarity, Malcolm distinguishes three degrees of interaction, which he identifies as social coexistence, religious syncretism and theological equivalentism. The discussion of syncretism links to the chapter by Badalanova in pointing out that, at least at the level of folk religion, many practices are shared between Muslims and Christians; certain shrines and religious buildings are used by both groups (as mentioned also by Norris in his introduction); and there are also some shared rituals, such as processions – which are probably in any case a pagan survival. Malcolm also reminds us of the role of theological arguments in religious life, while most modern studies tend to deal with religion as a social phenomenon, attributing conversion purely to social and economic factors. Here he uses the term 'theological equivalentism' to describe the argument that both Christianity and Islam are equally valid ways to salvation. He then turns to the main focus of his chapter and, basing his account on a detailed analysis of the earliest evidence, shows that crypto-Christianity proper emerged in the seventeenth century and continues to the twentieth. He suggests that while the practice may have evolved for reasons of expedience – to avoid paying certain taxes levied only on Christian men – many men may have converted to Islam, while remaining Catholics within their own homes. But he suggests that some recent instances must be evidence of something else: a genuine parallel commitment to two distinct ways of life.

Two chapters then consider aspects of Orthodoxy: Vladeta Janković describes the popular legends and beliefs associated with the most sacred icon in the Serbian Orthodox tradition. He focuses on Mount Athos as the central site of Orthodoxy and in particular on the Serbian monastery of Chilandariou, established in the twelfth century. The story of the icon of the 'Three-Handed Virgin', which is kept there, is intriguing in its own right and also provides an illustration of widespread beliefs in the miraculous powers of various sacred objects.

Muriel Heppell examines one important trend in Balkan Orthodoxy, Hesychasm, which developed in the course of the fourteenth century. She gives a detailed account of the life of one of the most important figures in the movement in the Balkans, pointing the way forward to future research. Her focus is Bulgaria, but by describing the spread of the movement also to Serbia and the Romanian principalities of Moldavia and Wallachia, she illustrates the unity of the Orthodox world at this period, when monks often travelled considerable distances between monasteries. Heppell concludes by suggesting parallels between Hesychasm and other religious trends both within the Western church and in other faiths.

The next three chapters concern Islam in the Balkans: offering first general discussions of its position in Bosnia and Romania, and then an account of the most influential Muslim order under Ottoman rule: the Bektashis. Alexander Lopasic begins by considering the large numbers of converts to Islam in Bosnia in the light of historical research. He includes a detailed account of the controversial 'devshirme' system, mentioned by Heppell in the Introduction. This was the practice whereby village children were taken from their homes in order to serve the Ottoman state, and particularly as recruits for the army. He describes also the effect of the Habsburg Military Frontier on the growth of Islamicisation, since it entailed the establishment of frontier units with commanders, a hereditary post, which led to the growth of a real Bosnian 'frontier aristocracy'. Lopasic suggests that because Bosnian Muslims were mostly local people who converted over the whole 400-year period of Ottoman rule, Islam remained more or less intact after the Ottoman departure from Bosnia in 1878. To complete his account, Lopasic briefly discusses the period of Austrian rule and the endeavours between 1882 and 1903 to foster the idea of a Bosniak nation: to include the whole population – Catholic and Orthodox Christians and Muslims. He ends his comprehensive account with a discussion of the development of Muslim identity within the state of Yugoslavia. Jennifer Scarce begins her description of Islam in the Dobruja region of Romania with an interesting new angle, showing that the Muslim community there predates the Ottoman conquest, having begun with the immigration of Tatars from the ninth century on. Muslims tend to be concentrated there since, after that early beginning, the only areas of Romania which were under direct Ottoman rule were the Banat in the north and the Black Sea region of Dobruja. Scarce gives an account of the particular ethnic mix of Muslims in the region and describes their way of life and accommodation with the state.

After these two survey chapters, John Norton focuses on one Muslim order – Bektashism – in much the same way as Heppell examined the Hesychast movement within the Orthodox Church. This chapter gives a detailed account of the basic tenets of the Sufi orders, to which the Bektashis belong; the origin of the order, its particular features, its connection with the Janissaries (which partly explains the prominence of the Bektashis in the Ottoman Empire); and its fate in more modern times. After this general description of Bektashism, Norton outlines the spread of the order in the Balkans, pointing out that it is particularly strong in Albania, where it survived the break-up of the Ottoman Empire more successfully than in the rest of the Balkans.

The last two chapters in the volume consider wider issues of the role of religion in the context of late nineteenth-century Bulgaria, in the last decades of Ottoman rule, on the one hand, and twentieth-century Bosnia on the other.

F. A. K. Yasamee considers the recently published diaries of the Bulgarian Exarch Yosif, dating from 1868 to 1915, which offer new insights into the affairs of the Bulgarian Orthodox Church and a personal view of the meaning of the significant events of the time. The chapter gives a fascinating account of the role of the church in raising Bulgarian national consciousness: as in other Balkan nations, Orthodoxy is seen as a defining element of Bulgarian national identity. Yasamee also analyses the involvement of the church in political events, such as the delicate relations between Bulgaria and Russia, and the rise of the Macedonian national movement. As has so often been the case in this volume, one detailed case-study may serve as a specific instance of general trends, applicable throughout the region.

Mitja Velikonja takes the example of Bosnia to analyse religious and mythological factors in creating a sense of national identity – in this case, the prevailing national mythologies of the component peoples of the territory. He sets out to cast light on the part played by such mythologies in the recent conflicts in the central Balkans. Velikonja suggests that 'the contemporary religious and national history of Bosnia and Hercegovina offers a perfect and tragic example of the extent and potency of the political abuse of religious-national myths.' The bulk of this volume has dealt for the most part with the role of religion in individual daily lives, the origins and development of specific religious communities and movements. This chapter takes a broader view of the way such 'raw material' may be exploited for particular political ends. The chapter clearly explains the function of mythology, which the author sees as existing in both pre-modern and contemporary societies. He examines the way the respective churches in Bosnia Herzegovina became involved in the formation of the key religious-national mythologies of each community, and the disastrous consequences that ensued.

In a reflection on the subject matter of this volume, a similar issue is raised by John Nandriş, the author of its first chapter. It concerns the exploitation by outside interests of the human capacity for belief. In his chapter, Nandriş warns us that we should not ignore the dark side of belief. That 'dark side' has been present throughout this volume, in references to the persecution of numerous different groups by others in the name of religion. In his longer reflection, Nandriş considers the impact of some more recent 'pseudo-religions', such as Communism. We include a passage from it here, as a fitting discussion for the end of this brief survey of the book's contents:

Communism crystallised Marxism as a pseudo-religious distortion of religion, in the totality of its claims and the demands that it made upon its subjects, in its utopian aspirations, and in the quasi-religious comforts which it extends as an opiate to the *literati*. The outcome of

Communism was the inevitable result of the Marxist algorithm contained within it, not least its rejection of monotheistic belief.

Authoritarian socialism may indeed have incorporated mandatory forms of common worship, sacred texts, the mummification of its saints, heresy, inquisition, iconography, or the confessional, but it lacked the ability to create sincere belief amongst its practitioners. The poverty-stricken philosophy of poverty was hardly a competent tool for economic analysis. What is most remarkable is that, despite strict indoctrination from birth, young people living under socialism proved to possess an intrinsic moral sense that enabled them decisively to reject its falsehoods.

Since nobody was a true believer in Communism, what belonged in theory to everybody belonged in practice to nobody. The contrast with Christian monasticism is very striking. The monasteries were perhaps the only, certainly the most truly successful and enduring, collectives in European history. They were motivated by religious belief. They constitute a major topic in the history of Balkan religion in which they were active centres of learning and suppliers of social services.

The pseudo-religion of authoritarian socialism was not intended as a benevolent structure, but as one of the most comprehensive systems yet evolved for maintaining control of society for the benefit of a small minority. The apartheid of the *nomenklatura* was not based on skin colour, but it was a form of socialist priesthood initiated into the nods and winks of the pseudo-religion.

By way of conclusion, I would like to highlight Nandriş's phrase about the intrinsic moral sense which enables people to reject falsehood. This is the other side of the coin of manipulation of religious-national mythologies, which has had such a destructive impact on the Balkans. Elsewhere, again, Nandriş refers to several Christian hermits re-establishing themselves in the remote Carpathians during the last two or three years of Ceausescu's regime. Based on the evidence of this volume, perhaps the most appropriate single word which may be said to characterise the history of religion in the Balkans is 'resilience'. This applies to individuals as much as to whole communities: at every stage in the evolution of the religious life of the Balkans, there have been instances of cynical exploitation and oppression, and people eager to follow the compelling call of demagogues and false prophets. This is one of belief's dark sides. But, equally, there have always been those whose moral courage is unwavering. The Appendix to this volume, describing a

meeting between representatives of the three main religious denominations in Kosovo (Kosova), in March 1999, on the eve of the NATO
bombing, offers a poignant instance of the potential of the genuinely
devout to reject falsehood and manipulation. For the time being, the
vision described in this text has been utterly lost. Nevertheless, there is a
great deal more to be said about the brighter side of the religious life of
the Balkans, and we can only hope that its indisputable capacity for
resilience, mutual respect and reconciliation will ultimately prevail, free
from the interference of power-hungry politicians. In the meantime, a
better understanding of the complex processes woven into the cultures
of this part of Europe is essential to ensure its more peaceful and pro-
ductive future. We hope that this volume will make a small contribution
to that understanding.

The editors wish particularly to thank Sonia Kanikova for her invaluable
work and to acknowledge the generous support of SOAS in letting us
use the Brunei Gallery for the conference itself; the Archbishop of
Canterbury; the Right Reverend Roger Sainsbury, Bishop of Barking;
Lady Marks; the British Association for Central and Eastern Europe;
the British Academy and the Elizabeth Barker Fund, without all of
whose help this project could not have come to fruition. We would
like also to thank the Director of SSEES, Professor Michael Branch and
Mr Tim Farmiloe of Palgrave for including this volume in the SSEES–
Palgrave publications. Our thanks go also to Catherine Pyke of the
Geography Department at UCL for the maps and to the editorial staff
of Palgrave for their work on the final stages of production.

CELIA HAWKESWORTH

Notes on the Contributors

Florentina Badalanova is a research associate at the School of Slavonic and East European Studies, where she taught Bulgarian language and literature, 1994–98. She has been engaged in research into popular culture since 1981. She is currently working on 'The Folk Bible' (Bulgarian Oral Tradition and Holy Writ), 'The Folk Koran' (The Book of Fate and the Living Faith) and a 'Folklore Dictionary of Saints in Slavia Orthodoxa'.

Bernard Hamilton is Professor Emeritus of Crusading History in the University of Nottingham. His recent publications include: with Janet Hamilton, *Christian Dualist Heresies in the Byzantine World c. 650–c. 1450*, MUP, 1998, and a volume of essays, *Crusaders, Cathars and the Holy Places*, Ashgate Publishers, 1999.

Celia Hawkesworth is Senior Lecturer in Serbian and Croatian Studies at the School of Slavonic and East European Studies. Her publications include *Ivo Andrić: Bridge Between East and West*, Athlone Press, 1977, *Voices in the Shadows: Women's Writing in Bosnia and Serbia*, CEU Press, 1999, and (ed.), *A History of Central European Women's Writing*, SSEES–Palgrave, 2001.

Muriel Heppell is Emeritus Reader in the Medieval History of Orthodox Eastern Europe in the University of London. She has travelled widely throughout the Balkan area and spent ten years teaching English in the University of Novi Sad. Publications relating to this area include articles on various aspects of monastic literature, and *The Ecclesiastical Career of Gregory Camblak*.

Vladeta Janković is Professor of Classical Literature at the University of Belgrade. His primary research interest is classical comedy, and the influence of religion on literary phenomena. His main publications include *Menander's Characters and the European Drama*, 1978, *The Comedies of Terence*, 1978, *The Laughing Animal – On Classical Comedy*, 1987, *The Drama of Hroswitha*, 1998 and *Myths and Legends – Hebrew, Christian, Islamic*, 1995.

Alexander Lopasic was Lecturer in Social Anthropology at the Department of Sociology, University of Reading. His primary research interest is Islam in Africa and the Mediterranean. His main publications are *Commissaire Général Dragutin Lerman (1863–1918): a Contribution to the History of Central Africa*, Tervuren, 1971 and (ed.), *Mediterranean Societies: Tradition and Change*, Zagreb, 1994.

Noel Malcolm was Fellow of Gonville and Caius College, Cambridge, from 1981 to 1988 and later became Foreign Editor of the *Spectator* and political columnist on the *Daily Telegraph*. His publications include *De Dominis, 1560–1624: Venetian, Anglican, Ecumenist and Relapsed Heretic*, *George Eminescu: His Life and Music* and two major works on the Balkans: *Bosnia: a Short History*, 1994 and *Kosovo: a Short History*, 1998. He is currently working on a biography of Thomas Hobbes.

Elissaveta Moussakova, PhD in Art History, is Head of the Department of Manuscripts and Old Printed Books at the National Library, Sofia. She works and publishes in the field of illuminated Bulgarian medieval manuscripts, Bulgarian medieval art, signs and symbols in Christian art. She is Visiting Lecturer at the New Bulgarian University, Academy of Fine Arts (Sofia) and Central European University (Budapest).

John Nandriş FSA was until recently Senior Lecturer in European Prehistory at the Institute of Archaeology in University College London. He has undertaken field work in every country of Europe, and published in some dozen of them on subjects as various as flotation for prehistoric plant remains, the neutron activation analysis of prehistoric obsidian, the Neolithic of Greece, the replication of artefacts, the culture of the Aromani, the Jebaliyeh Bedouin of Sinai, or the ethnoarchaeology of Daghestan. His research interests centre in South-East Europe, and most recently in the ethnoarchaeology of the highland zone which has taken him further afield, to the Atlas, Sinai, the Caucasus, and even Japan.

Harry Norris is Professor Emeritus of Arabic and Islamic Studies, University of London. His research and publications have spanned the Arab world, but his recent studies have concerned the Muslim communities of Eastern Europe. His recent publications include *Islam in the Balkans*, Hurst, 1993; a contribution to *Bosnia, Destruction of a Nation, Inversion of a Principle*, Islamic World Report, 1996 and, as co-editor, *The Changing Shape of the Balkans*, UCL Press, 1996.

John D. Norton is Director of the Centre for Turkish Studies, University of Durham. He has held additional posts in the university as Director of the Centre for Middle Eastern and Islamic Studies and Principal of St Cuthbert's Society. His research interests include Turkish religious movements, modern Turkish history, Cyprus and Turkish painting.

Ioan-Aurel Pop is Professor of Medieval and Early Modern History at the 'Babeş-Bolyai' University, Cluj-Napoca, Romania. His main publications include *Romanian Medieval Institutions*, Cluj-Napoca, 1991, *Romanians and Hungarians from the 9th to the 14th Century. The Genesis of the Transylvanian Medieval State*, Cluj-Napoca, 1996, *The Medieval Genesis of the Modern Nations*, Bucharest, 1998 (English) and *The Romanian Medieval Nation*, Bucharest, 1998.

Jennifer Scarce was Curator of Middle Eastern Cultures, National Museums of Scotland and is now Honorary Fellow, Department of Islamic and Middle Eastern Studies, University of Edinburgh. She has travelled extensively in Iran, Turkey, the former Ottoman provinces in South-East Europe and the Arab Gulf and has published widely on the region. Her current research interest is the impact of Islam on Romania in terms of settlement patterns and material culture.

Mitja Velikonja is Assistant Professor at the Faculty of Social Sciences, University of Ljubljana and Research Fellow of the Centre for Cultural and Religious Studies, there. His publications on contemporary mythologies and religious-cultural dynamics in Central and Eastern Europe include *Masadas of the Mind: Crossroads of Contemporary Mythologies*, Ljubljana, 1996 and *Bosnian Religious Mosaics: Religions and National Mythologies in the History of Bosnia Herzegovina*, Ljubljana, 1998.

F. A. K. Yasamee is Senior Lecturer in the Department of Middle Eastern Studies, University of Manchester. His main research interests are the history of the Middle East and South-Eastern Europe in the nineteenth and twentieth centuries. His recent publications include *Ottoman Diplomacy: Abdülhamid II and the Berlin Settlement 1878–1888*; 'The Ottoman Empire' in K. M. Wilson (ed.), *Decisions for War 1914*; 'Nationality in the Balkans: the Case of Macedonia' in G. Ozdogan and K. Saybasili (eds), *Balkans: a Mirror of the New International Order*; and 'Colmar Freiherr von der Goltz and the Rebirth of the Ottoman Empire' in *Diplomacy and Statecraft*, 9(2), 1998.

Map 1 East Central Europe, ca. 1930

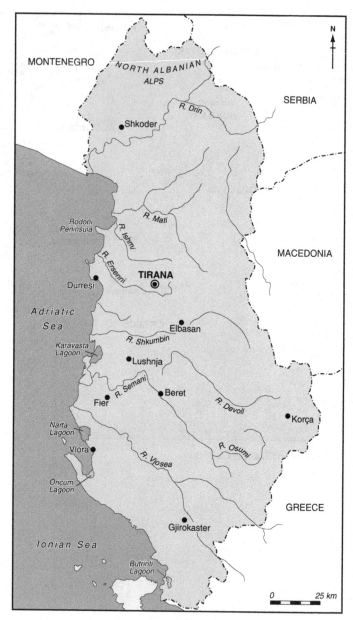

N

MONTENEGRO

NORTH ALBANIAN
ALPS

SERBIA

R. Drin

● Shkoder

Rodoni
Peninsula

R. Mati

R. Ishmi

R. Ersenni

MACEDONIA

TIRANA
◉

● Durresi

Adriatic
Sea

Elbasan ●

R. Shkumbin

Karavasta
Lagoon

● Lushnja

R. Semani

Narta
Lagoon

Fier ●

● Beret

R. Devoll

● Korça

● Vlora

R. Osumi

Oricum
Lagoon

R. Vjosea

GREECE

Ionian Sea

● Gjirokaster

Butrinti
Lagoon

0   25 km

Map 2 Albania

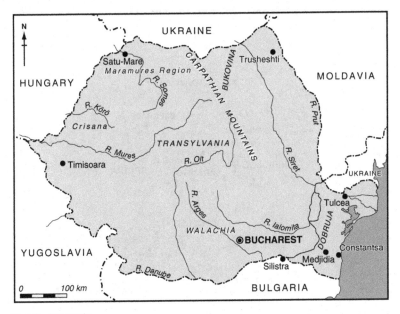

Map 3 Romania

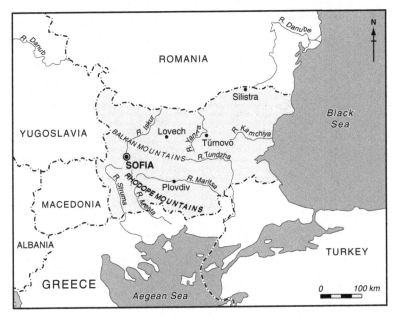

Map 4 Bulgaria

Map 5   Yugoslavia, 1992

Introduction

Muriel Heppell and Harry Norris

Balkan Christianity

Muriel Heppell

At the beginning of the fourth century AD, when the Roman Emperor Constantine established a separate capital (named after himself) for the eastern half of the empire, Christian churches had been established throughout imperial territory; and this included the area of the Balkans, south of the river Danube. Most of these churches were organised under the leadership of local bishops; there were, however, three ecclesiastical centres of more than local importance: Rome, Alexandria and Antioch; and later Constantinople, the eastern capital.

During the fourth and fifth centuries, much of the western empire was overrun by pagan barbarian tribes which eventually settled on the territory they had invaded. As a result, many parts of the former empire (including the Roman province of Britain) became, in effect, missionary areas. Somewhat later this was the situation in the Balkans, most of which had been settled by different groups of Slav tribes by the end of the seventh century. But the Balkans, unlike the missionary areas of northern and western Europe, were visited by missionaries from both the eastern and the western centres of the Christian faith, that is from both Rome and Constantinople. Although there was no formal division within the Christian Church until the schism of 1054, there were already differences between the churches of the eastern and the western parts of the former Roman Empire with regard to the form of public worship (including the liturgical language), certain customs, and in the organisation of monastic life. These differences were reflected in the religious life of the Balkans, since during the eighth and ninth centuries the Slovenes and Croats who had settled in the north, and

1

the Magyars (who had settled in what later became the kingdom of Hungary) became part of the western, Catholic Church, while the Bulgarians, Serbs and Macedonians became part of the Eastern Christian, Orthodox Church. This difference had lasting effects on all parts of the region, not only in their religious life, but also in their cultural and political development.

From the time of the conversion to Christianity, the religious life of the Balkans showed distinctive features, some of which are illustrated in the contents of this volume. One of the earliest of these was the introduction of a new liturgical language, in addition to Greek and Latin. This was the work of two Macedonian monks, the brothers Constantine and Methodius, Byzantine Greeks by education, but probably of mixed Greek/Slav descent. The immediate pretext for the introduction of this language was a request to the Byzantine Emperor Michael III, in 863, from a Western Slav prince, Rastislav of Moravia, asking the emperor to send him missionaries who could instruct his subjects in the Christian faith in a language they understood. At this time, the vernacular languages of the Slav peoples throughout the Balkans, from Macedonia as far north as Moravia, were sufficiently similar to be mutually comprehensible. Moravia was already officially Christian, under Roman jurisdiction; but the new religion was making slow progress because the church services were in Latin, which the people could not understand. The emperor turned for help to the monk Constantine, who already had experience of missionary work, and was known to possess exceptional linguistic gifts.[1] He then undertook the task of developing a Slav script, based on phonetic principles. The resulting alphabet came to be known as glagolitic. This was later modified, and the final Slavonic alphabet, incorporating several Greek letters, is called Cyrillic, from the name Constantine adopted after his final monastic profession, just before his death in Rome in 869.[2]

Constantine then translated the liturgy and parts of the gospels into the vernacular language of his native Macedonia, which came to be known as Old Church Slavonic, and set out for Moravia with his brother Methodius. In spite of papal approval of their work, they soon incurred the hostility of the German bishops in Moravia, and eventually (after the death of Constantine), Methodius and his disciples were forced to leave. Later, after Methodius died, his disciples made their way to Croatia, where the Slav liturgy was at first enthusiastically received, though later it was banned.[3] It was not until the conversion of Bulgaria in the ninth century that the full potential of Old Church Slavonic as a liturgical and literary language developed. This owed much to the valuable pioneer work of the Bulgarian monks Kliment (Clement) and Naum, and their disciples.[4] From Bulgaria this increasingly formal Church language passed to the Serbs and later to the Russians, and also to the Romanians (who were Orthodox, though not Slav). In the course of time a large volume of both translated and original literature appeared in this language.

Another early feature of Balkan religious life, after the acceptance of Christianity, was the emergence of a particularly uncompromising form of dualist heresy known as 'Bogomilism', after the name of its best-known leader, a priest named Bogomil (Theophilus). This also appeared first in Bulgaria, in the tenth century, although it soon had adherents in other parts of the Balkans, notably in Serbia, and also in the Byzantine Empire. The Bogomils condemned and rejected everything of material origin, since they considered such things to be derived from an evil creative force, a fallen angel hostile to God, the primary source of creation.[5] They also rejected all forms of secular and ecclesiastical authority, as being of human origin. As a result, these authorities made every effort to punish the Bogomils and root out their heresy. However, in spite of persistent and sometimes brutal punitive measures, Bogomilism was never completely eradicated, although it was driven underground for a time. It enjoyed a considerable revival in Bulgaria in the thirteenth and fourteenth centuries, and also flourished in Bosnia in the later Middle Ages.

It was long believed that there was a close connection between the Bogomil heresy and the emergence, in the thirteenth century, of the organisation known as the 'Bosnian Church'. This certainly had a hierarchical structure similar to that of the earlier Bulgarian Bogomil groups, and there is some evidence that the beliefs of the Bosnian Church included some Bogomil elements. One way in which it differed from the earlier Bogomil movement, however, was that it had more broadly based social support, since its adherents included landowners and members of the nobility, as well as peasants and poorer members of society.[6] The traditional view of the Bosnian Church as a basically Bogomil organisation has been challenged by John Fine, who considered the members of the Bosnian Church to have been schismatic Catholics.[7] Actually, these interpretations are not incompatible, since the members of the Bosnian Church may well have considered themselves to be Catholics, while at the same time advocating a less extreme form of dualism than that preached by the earlier Bogomils. Dualism, like Protestantism in western Europe in later centuries, was not necessarily a fixed or uniform movement, and could have taken different forms.

In the year 1054, after mutual excommunication, and recriminations, the Orthodox and Catholic branches of the Christian Church entered a state of schism which has lasted to the present day, in spite of attempts to heal it. This naturally affected the Balkan area, since the northern inhabitants had been converted by western missionaries, while those of the south and east, apart from the Adriatic coast, received their Christianity from Byzantium. In other parts of Europe, relations between the separated Churches gradually hardened into permanent attitudes of mutual hostility. In the Balkans the situation was less clear cut. For example, both King Kaloyan of Bulgaria (1197–1207) and Stephen the First-Crowned (ruler of Serbia from 1197 to 1227) sought royal insignia from Rome.[8] During the thirteenth century Serbian rulers made dynastic marriages with Catholic as well as Orthodox ruling families, and in general maintained friendly relations with Catholics within their dominions and on their borders, although commitment to Orthodoxy as the national religion of the Serbs never wavered; as one writer has put it: 'The Eastern . . . was the "established" and the Western the "disestablished" church in the Serbian state.'[9] It is true that a harsher note emerges in the legal code ('Zakonik'), promulgated by the Serbian ruler Tsar Dušan in 1349: heretics (presumably mainly Bogomils) were to be branded and expelled from the country (Article 10); proselytism was forbidden (Article 8); on the other hand, foreign merchants travelling through the tsar's dominions were not to be robbed or molested in any way (Article 112). It is true that these apparently more tolerant attitudes were motivated by political and economic self-interest rather than by any abstract spirit of enlightenment; but it is interesting to note that the situation in the Balkans made such attitudes possible.

From the early sixteenth century onwards, most of the Balkan area became part of the Ottoman Empire, whose religion was Islam, although the northern part of this area (Hungary and Croatia) did not remain under Ottoman rule for very long.[10] But the inhabitants of the central and southern Balkans remained under Muslim rule for at least 400 years, and longer in some cases. The Ottoman rulers did not proscribe the Christian religion or persecute their Christian subjects: one element in the Ottoman administrative system was based on units defined by religious allegiance rather than ethnic identity. Hence all the Balkan Christians – Bulgarians, Serbs, Bosnians and Greeks – formed a Christian 'millet'. However, the Christians were subject to a special tax ('harač') and to the so-called 'tribute of blood' (devshirme): the forcible deportation of young males who were converted to Islam and trained for service in the Ottoman army or administration.

In spite of the impoverished condition of many of its adherents, the life of the Orthodox Church in the Balkans continued and even flourished at times, especially after the revival of the Serbian Patriarchate at Peć in 1557.[11] During the sixteenth century several new churches were built and monasteries founded, especially in Bosnia and Herzegovina.[12] This is particularly interesting, since Bosnia was the only region where conversions to Islam after the Turkish conquest were at all numerous. At one time it was thought that the Bosnian converts were members of the Bosnian Church, for whom it was easier than for other Christians to accept the strict monotheism of Islam. However, recent research has established that in fact conversions to Islam came from all religious groups, and that the process of 'Islamicisation' in Bosnia was spread over a long period of time.[13] The most likely reason for the greater number of conversions in Bosnia is the fact that neither of the medieval Christian Churches, Catholic or Orthodox, had ever had a really strong administrative structure there. And because ecclesiastical authority had always been weak, it was easier for individuals to make their own religious arrangements[14] – as may be seen from some of the contributions in this volume.

The story continues. In the second half of the present century, most of the religious groups in the Balkans – Catholic, Protestant, Orthodox and Muslim – had to adapt themselves to living under Communist governments whose ideology was fundamentally unsympathetic to their religious beliefs. Since the collapse of Communism, they have faced the difficult task of reviving their spiritual life in a predominantly secular society.[15] Both processes have been well documented, and will no doubt be chronicled and interpreted in detail by future historians.

Balkan Islam

Harry Norris

The Balkans represent a distinct region within the world of Islam. Balkan Islam possesses a unique identity and it has given birth to distinguished personalities within its three chief centres, Bosnia and Herzegovina, the Albanian (Arnaut) territories – Albania proper, Kosovo/a and Western Macedonia – and Bulgaria. These personalities include: Ottoman governors, Koran commentators, men of state such as Muhammad Ali, in Egypt, and poets of merit, such as the Albanian Bektashi, Naim Frashëri.

In the early medieval period, two areas of Europe were extensively Islamised: the Iberian peninsula and the Central Volga region, together with parts of the Northern Caucasus. It was in the later medieval period that the Balkans, together with Hungary and the Crimean peninsula, began to be Islamised under direct and indirect Ottoman rule. Culturally, the Iberian peninsula was linked to North Africa, the Volga region to Central Asia. The Balkan peninsula was an Ottoman fiefdom and a frontier district. In addition to its connection with Ottoman Turkey, it was linked in faith and in culture with the Arab Middle East, especially with Egypt and with Syria. It was also extensively influenced by Iranian thought and material culture, intellectually in the case of part-Shiite Albania and culturally in the case of pre-Muslim Bulgaria.

Whilst India, East Asia and Africa have marked the shifting border between *Dar al-Islam* and the non-monotheistic world of *Dar al-Kufr*, the Islamic communities of the Balkans – as is also the case in Russia and the Caucasus region – have coexisted with, contended with and fought bitterly with their Christian compatriots, by birth or by adoption: compatriots who frequently shared a common language and who were either professed Catholics or Eastern Orthodox. They have done this for centuries, and it seems inevitable that they will continue to coexist in such a way in an uncertain future, and at a grave disadvantage, into the next millennium. Within its own belief systems, Islam in the Balkans has woven its multicoloured *kilim* over a thick underlay of Christian and pre-Christian belief. The presence of these beliefs, very often thinly disguised, may be detected in every Balkan Muslim community. It has been so in all levels of society and in all ages since the faith of the Prophet established a foothold and subsequently put down roots within South-Eastern Europe.

The Balkan peoples are predominantly Eastern Orthodox. Islam is frequently far from Orthodox, if the latter is assumed to be synonymous with the Hanafite Sunnism of the Turkish and Arab faithful in the Middle East. The Balkan Muslims, and especially the Bosniaks, claim that their Islam is the most representative, today, of what they maintain is 'European Islam', and this despite profound differences between Bosnians and Albanians. Religion is a way of life, an attitude of mind and a mark of identity rather than a creed or a dogma. Moral values and standards take precedence over confession. The Bosnian Muslims' claim to be heirs to an ancient Balkan tradition, which is deeply rooted, is nowhere more sharply defined than in the creed of the mystics who are members of the Sufi brotherhoods (*tarikat*). According to Rusmir Ćehajić:

The spiritual genealogy of the Bosnian Sufi proceeds from Jesus, the son of Mary, to Muhammad, son of Abd-Allah, closing a full circle, surrounded by rose petals and vine branches, the two basic symbols of Bosniak spirituality. The circle is composed of those carrying over the living tradition from Jesus to a line of patriarchs of the Bosnian Church, who handed their spiritual trust to Sufi Shaykhs, who in turn received their knowledge through spiritual genealogy from the Prophet, recipient of revelation from God through the Archangel Gabriel.[16]

The above confession brings to the fore the inherent tendency in Balkan Islam to draw upon spiritual power and solace from the faiths which preceded it. At times, be it in festivities, shared calendars or blessings from Christian clergy, this has taken place when the clergy in question have been under pressure from the Muslim community surrounding them. In Istanbul, for example, it has been reported by Western journalists and visitors that barren Turkish women visit Orthodox churches in order to drink Holy Water in the hope of some cure for their infertility. Newly married couples may also ask Orthodox priests for a blessing, deeming this to be more effective and of greater spiritual power than that of their *imams*.

The Bosniaks claim that their faith is a 'European Islam', an organic growth from past religious beliefs and not a transplant like the Asian Islam which has recently arrived in the West. This case was strongly made by Professor Enes Karić of the Faculty of Islamic Studies in Sarajevo in a recent conference paper. Refuting what he terms 'the fascist theory', which assigns Islam in Bosnia to the position and status of an alien and imported faith, he remarks:

I, Muslim by faith and Bosniak from Bosnia by ethnic background, am telling you here and now, that when it comes to religion, Europe shares the destiny of its ancient religious founder and spiritual mainstream, Asia. That means that Europe, just like Asia, is the continent of Islam, as well as of Orthodox Christianity, and of Catholicism, and in the future, most probably, of Buddhism. The leaf must come after the branch, the branch must come after the tree, the tree must come after its roots![17]

In other circles of Balkan Islam, much more legalistic, even 'fundamentalist', tendencies may also be observed. Yet others are by nature austere and puritanical, or, yet again, they may be almost wholly secular in their outlook. Pride in the ethnic definition of 'Muslim', sharpened and clarified by the war in Bosnia, is a label which may have little regard to firmness of conviction as to the validity of Islam's five pillars of faith. At times, a vague spirituality, open to ideas which originate in India or in the New Age sects, has tempered dogmatic assertions. This recalls a remark once attributed to the composer, Mauricio Kagel, who, when asked whether composers nowadays believed in God, expressed the opinion that 'few believed in God, but that all of them believed in Johann Sebastian Bach'. Describing Bosnia in the mid-1980s, Dr Cornelia Sorabji emphasises that:

> Very few went so far as to deny outright the existence of God and the validity of the Prophet. The consumption of alcohol and other behaviours so often used in evidence that the Muslims are not 'real' Muslims weighs as little as the thefts committed throughout the Christian world. For example, because a Catholic steals, one would not say that s/he was not a Catholic, but simply that s/he had broken one of the rules of the Catholic faith.
>
> This is an important point, for, like the other religions of the region, Islam is not simply a set of clearly definable rules of practice and it is not so understood by Muslims. Alongside the detailed prescription for daily prayer, the annual Hajj/Hadji/Hajji, and so on, Islam is also understood as a domain of loose moral imperatives – hospitality, cleanliness, generosity, honesty, kindness, courtesy, industry and so on. The first three are often regarded by Muslims as virtues especially belonging to Islam, but all form part of a general field of morality that can potentially be seen as overlapping with that of non-Islamic ideologies or non-Bosnian Muslim societies.'[18]

Islam in Bosnia is, arguably, the most academically impressive and has been the most influential because of its scholars and the wealth of its vast Islamic literature. Libraries once graced major cultural centres, such as Sarajevo, Mostar, Travnik, Prušac and Zvornik, as well as numerous Sufi *tekkes* and the private libraries of wealthy Bosniaks. The historical, social and economic reasons for this are presented and discussed by Dr Alexander Lopasic, in his contribution 'Islam in the Balkans: the Bosnian Case'.

Islam in the Albanian-speaking communities in the Balkans may appear to be markedly different. It is here (with the exception of the Alawite communities, especially in Deli Orman, in Bulgaria) that the influence of Shiite Islam is most clearly apparent.[19] Amongst the

Albanians, many pre-Islamic ('Illyrian') customs and beliefs survive. Several of these were described and explained in Dr Elira Lamani's conference paper, entitled 'Diffusion of Islam in the Life of the Albanians during Ottoman Rule':

> While spreading in all aspects of Albanian social life, Islamic influence encountered the ancient cultural traditions and customs of the people. In this respect, the historically unavoidable influence of Islamic culture has had special features. It began to be diffused mostly during the seventeenth century, when conversions increased the number of Islamic believers, especially among the Catholics of the North. From the report of a Catholic missionary on 27 March 1638, we learn that 'in some of their customs Albanians are so stubborn that they would rather be killed than give them up. In the mountains, they would celebrate feasts according to their customs, and would appreciate theirs more than those of the churches and mosques.' Among the elements of nature which they worshipped, the sun occupied the main place. In studying the customs of Albanians, the English traveller, Edith Durham, remarks that the cult of the sun was deeply rooted.[20]
>
> This attitude towards ancient customs or traditions is reflected in the way the people preserved the names of pre-Christian or pre-Islamic origin. Widely used proper names bear historic, social or natural references: for example names related to animals or birds (*Shpend*, bird; *Uke*, wolf, etc.); to plants and trees (*Lule*, flower; *Shege*, pomegranate, etc.); or to admirable or acceptable qualities (*Bardha*, white; *Misoshe*, good, etc.). These names have been preserved with love and fanaticism from being replaced by Christian or Islamic names. The preservation and the existence of these ancient Albanian traditional names, which were very frequent among people of various religions, was connected to the belief that those names had a protective function, affecting the health and life of the individuals bearing them. In some families, when babies died soon after they were born, they would give the next children born names such as *Guri* (stone), *Mali* (mountain), *Rrapi* (a solid and long-living tree), believing that the children bearing those names would be as strong as the stone, live as long as the mountain, etc.

It is hardly surprising that Balkan Islam has been described by writers, by researchers, by some archaeologists and anthropologists and others who have been enthralled by religious survivals, as a classic example of the religion of a people being shaped by heterodoxy and by popular superstitions. The heterodoxy in question may be either theosophical or philosophical, as I discussed in my own conference paper, 'Aspects of Islamic Antinomianism, Heterodoxy and Persian Monism in the Literature of the Muslim Communities of the Eastern Balkans'. It owes much to poetic thinkers and religious mavericks in the Middle East and beyond, but it may also be a 'home-grown' affair, originating in cults which were present in the Balkans long before the mass conversion of its peoples to varied forms of Christianity and Islam.

This heterodoxy has been well documented over many years,[21] but far less is known about the survival of crypto-Christianity in regions where Muslims are now the majority, such as parts of Bulgaria, and in Kosovo, discussed by Noel Malcolm in this volume. Indeed, it may well be possible that the survival of so-called 'heterodox' Christian congregations, who are in fact 'crypto-Christians', may, in some cases, explain the origin and even the survival of existing, allegedly 'heterodox' Muslim communities. In the Bulgarian Orthodox communities there are two Bibles: the canonical text of Scripture and a Folk Bible, which varies from parish to parish and even from one household to another.[22] Much the same is characteristic of the Muslim communities, where the Koran, in its canonical Arabic text, is all but unintelligible to many congregations and where its examples of Prophets (shared with Jews and Christians) and Biblical (or pseudo-Biblical) narrative are attired in the raiments of folklore, a folklore which is shared mutually by the laity whatever their professed religious allegiance might happen to be.

Throughout history, there has been an uneasy relationship between Muslim believers in the Balkans and their fellow believers in parts of the Middle East and the Muslim heartland. The regional problems of the Balkans have only recently become a major concern to the latter. While in matters of faith, in law and the Meccan pilgrimage, Balkan Muslims have turned to the Arab world, the literary, intellectual and speculative tendencies and inclinations of Balkan Islam have been nourished (often through Turkey and the Turks) by Iranian Islamic speculation. An alleged common 'Indo-European' mentality has been aired, from time to time, despite major differences which divide them. The atheist Enver Hoxha professed to have a profound admiration for the legacy of the Persians:

> The people of Persia have ancient progressive traditions, great culture and an extensive idealist philosophy. Writers, poets, philosophers and scientists who have astonished the world have emerged from their ranks. Even today, their works carry authority in the great world

treasury of culture.

The history of the Persian people and their outstanding representatives is one of the most glorious parts of world history. Many of these great men, philosophers and poets, such as Sa'adi, Ferdousi, Omar Khayyam, etc., etc., were from the common people and their writings had their source in the people, notwithstanding that they were supported by the Shahs of various empires. The tradition of this knowledge, of this science, has been handed down from generation to generation.[23]

Despite his Marxism, based upon his Bektashi background (coloured as it was by 'Hurufi' influences), Enver Hoxha perceived some merit in that Islamic stream which derived its inspiration, he alleged, from both India and Christianity in the Sufi mystic tradition:

Islam has also had its false prophets, but had their movements, for example that of the Motazilites in the 8th and 9th century (followers of a Moslem theological school which sought to bring together supporters and opponents of Ali), not been suppressed whenever they were created, they would not have permitted Islam to have had that spiritual development which it had later.

In this direction the *Sufi* movement was a movement of poets and mystics (they took their name from the white woollen robe – *suf*, wool – which they wore).

The mystical and ascetic current of Sufism presents itself as a special line of alleged perfection and saintliness of the believer, who could divorce himself from the world and ascend to Heaven by means of ecstasy. The Sufis abjured worldly blessings in order to get closer to paradise, to become the *muqaribun*, those appointed to God to be closest to him.

The members of this movement of dervishes formed secret societies to which they were admitted after a special preparation and indoctrination through prayers, mysterious movements and dances which were called *ziker*.

The Sufi sect of dervishes comprised: Mevlavi dervishes; Rufai dervishes; Bektashi dervishes.

It is claimed that Sufism was influenced by Indian (Buddhist and Vedic) and Christian mysticism.[24]

Enver Hoxha's sympathy with the highest flights of Persian idealism, despite his Marxism, may seem to be a strange contradiction. How could

Marx be squared with the Persian monistic belief in Man – and the physical form of Man, at that? And how could that belief be reconciled with dialectical materialism on the one hand and the immanence of the Divinity in every atom in the Universe on the other? The Persian 'Hurufi' philosophy harks back to Fadlallah of Astarabad, mystic, cabalistic, martyr and confessed incarnation of Deity, who lived in the fourteenth century. This school surfaced in the ideas and ideals of a number of Balkan Sufi orders, but none so noticeably as in the Bektashiyya brotherhood in Albania and Kosovo. It is at the heart of much of the verse of Naim Frashëri, Albania's most famous poet. This curiosity would appear to be less curious if one examines Hurufism's appeal outside the Balkans. Hamid Algar has drawn attention to this, in his article on the influence of Hurufism on Bektashism, which includes reference to Albania: 'Influenced by Soviet writings that extol the Hurufis as humanists, materialists and strugglers against feudalism, some Iranian Marxists began in the 1970s to adopt Fazlullah Astarabadi and Nesimi as their heroes.... Some Shii writers responded with an equally revisionist view of the Hurufis as representatives of "authentic" revolutionary Shiism.'[25]

The subject of Persian influences, whether poetic, religious or philosophical, on Albanian thought, raised *en passant* by Enver Hoxha, is discussed in far greater depth in a little known article, printed in Marxist Albania, by Zia Xholi. To cite an English résumé (by an unknown author, conveyed to me by Bejtullah Destani), Zia Xholi's views of Naim's philosophy from a Marxist premise were the following:

> The most eminent 'Enlightener' of that period was Naim Frashëri (1846–1900), born into the family of an impoverished country squire. He studied at the Greek Lyceum of Janina where he made his first acquaintance with the Greek and Roman classics as well as the French writers of the XVIII century. At the same time, Naim Frashëri showed a lively interest in the words of Persian poets and thinkers. Thus, his ideological formation was based on three sources.
>
> Naim Frashëri took the initiative in publishing the first Albanian review, 'Drita' (The Light). He was also instrumental in the opening of the first Albanian school in 1887 in Korcha. He himself provided the text books. He maintained close contact with the Albanian intellectuals who had formed and were leading literary societies in countries such as Romania and Bulgaria.

But Naim Frashëri has entered Albanian history as a national poet. The 1880s were the period of his maturity as a creative artist. He had two sets of works published during that decade, two volumes of lyric poems: *The Flowers of Spring* and *Peasants and Shepherds*. Then came two monumental historic poems: 'Qerbela', describing the struggle between the descendants of Mahomet and the usurper Muawiya, and the 'History of Scanderbeg', the national hero of the Albanians who fought against the Turks in the fifteenth century. Naim Frashëri died in Istanbul in 1900, after a long illness.

The article continues with a long analytical study of Naim Frashëri's poetry and philosophical ideas. The author is guided by the tenets of Marxist-Leninist theory in his appreciation of his subject.

Naim Frashëri's conception of the Universe and of Man is essentially idealistic. He believes in God as the creator of all things. The same Universal Spirit rules the stars of the firmament and the life of men on earth. But Naim Frashëri's mental evolution takes him away from orthodox Muslim teaching towards a pantheistic conception of God. He thinks of Him not as a separate Supreme Being, but as part of Nature itself, in which He acts as an immanent driving force. God is present in all things that we see: the blossoming rose in the garden, the singing nightingale in a tree. God's church is not a particular building, but the infinite space filled with numberless celestial bodies that surrounds us. In the author's Marxist view, Naim Frashëri's philosophical conception was progressive as opposed to the orthodox reactionary Muslim religion. His passionate love of nature makes him the best propagandist who awakens patriotic feelings by singing the beauties of his homeland in his poems. He makes the Albanians feel proud of their country and ready to make sacrifices for its sake. By identifying God with Nature, the poet comes near to a conception of the world based on reason and the discoveries of science which embodies the ideas of the French philosophers of the eighteenth century. But Naim Frashëri came to his pantheistic philosophy mainly through the teaching of the Persian religious scholars of the Sh'i sect, which was introduced to Albania in the guise of Bektashism.

The author goes on to explain why Naim Frashëri could not emancipate himself completely from any religious belief and become a free thinker. He explains that he was a product of his time and that his thinking was conditioned by his environment. That is why he could not embrace a materialistic and atheistic philosophy of life and history. The poet seeks to create among his fellow-countrymen the enthralling harmony he sees in nature. He tries to explain natural phenomena without the help of God and thus becomes a forerunner of the thoroughly developed materialistic view. In his book entitled *Knowledge*, he gives an explanation of the Universe based on the discoveries of modern scientists from Giordana Bruno to Newton. He follows Charles Darwin in his theory on the origin of species and particularly the appearance of man, to the religious conception of an immutable world order created by God. Naim Frashëri opposes the scientific view of a world as progressing and going constantly forward, according to the newly formulated laws of the natural sciences. His notion of the world at large was closely connected with the immediate task of the Albanian people to liberate themselves from the yoke of Ottoman rule.

Naim Frashëri seems to displease the author with his belief in an immortal soul which returns to God after the death of the body. The poet commits an even more unforgivable sin from the point of view of dialectical materialism by believing in Heaven and Hell. In some of his writings, Naim Frashëri has given a modified view of this matter which is his belief in metempsychosis. As he puts it in a celebrated poem: 'I (the soul) a star, a bird, a ram, and also a human being more than once.'

Coming to the problem of conscience and other spiritual phenomena, the author notes with satisfaction that Naim Frashëri represents them as a particular manifestation of the working of our brain. This leads him to his next point which is that the purpose of acquiring knowledge is to extend the power of man over nature, and to form a rational view of the things that surround us.

Once again, Islamic 'heterodoxy' displays a leavening and a dynamic impulse in the Islam of the Balkans, both at a popular and at an intellectual and spiritual level. The variety, the influences of a cultural kind which it brought with it from the Orient and disseminated deep within South-Eastern Europe cannot be assessed by a simple study of local communities, however profound that study. Massaged statistics alone cannot determine our measure of Islam, our reading of its breadth and depth amongst Balkan peoples who profess to follow the faith of the Prophet.

Notes

1 See A. P. Vlasto, *The Entry of the Slavs into Christendom*, Cambridge, 1970, pp. 34–6.
2 *Ibid.*, pp. 37–9.

3 M. Heppell and F. B. Singleton, *Yugoslavia*, London, 1961, p. 70; Vlasto, *Entry*, pp. 196–201.

4 D. Obolensky, *The Byzantine Commonwealth*, London, 1971, pp. 96, 103–4; Vlasto, *Entry*, pp. 168–70.

5 See D. Obolensky, *The Bogomils. A Study in Balkan Neo-Manichaeism*, Cambridge, 1948; Vlasto, *Entry*, pp. 227–33.

6 Heppell and Singleton, *Yugoslavia*, p. 49.

7 John V. A. Fine Jr, *The Bosnian Church. A New Interpretation*, East European Monographs, Boulder and New York, 1975; John V. A. Fine and Robert J. Donia, *Bosnia and Hercegovina: a Tradition Betrayed*, London, 1994, pp. 22–6.

8 Vlasto, *Entry*, pp. 185, 217.

9 *Ibid.*, p. 221.

10 Heppell and Singleton, *Yugoslavia*, pp. 78–81.

11 Dušan T. Bataković, *The Serbs of Bosnia and Hercegovina*, Guernes par Limay, 1996, p. 29. It is interesting to note that the Serbian Patriarchate was restored thanks to the initiative of the Ottoman vizier Mehmed Sokolović, deported from Serbia as part of the 'tribute of blood'; his cousin, Makarije Sokolović, was appointed Patriarch.

12 *Ibid.*, pp. 30–2.

13 *Ibid.*, pp. 25–7; Fine and Donia, *Bosnia and Hercegovina*, pp. 35–7.

14 Fine and Donia, *Bosnia and Hercegovina*, pp. 42–4.

15 One of the papers presented at the conference, by Professor Georgi Kariev and Petio P. Petkov, indicated that there are signs of a revival of interest in Bulgaria in 'Palamite' or Hesychast spirituality, which flourished in Bulgaria in the fourteenth century.

16 Islamic World Report, *Bosnia: Destruction of a Nation, Inversion of a Principle*, London, 1996, pp. 53–4. This statement may be viewed as both a cultural and a religious affirmation of Bosniak ethnic identity. The rose and the vine are Christian symbols, so, likewise, the rose and the grape are commonplace in Sufi imagery, especially in verse. The Ottoman poet, Lamii, in his great 'rose ode' extols the martyr al-Hallaj: 'For a century it has given sweet fragrance to the soul, like the breath that quickened Jesus – like the narcissus, it captures the eye of discerning people, the Rose', see L. Massignon, *The Passion of al-Hallaj, Martyr of Islam*, Princeton, 1982, vol. 2, *Survival*, p. 256.

     The Bosniaks see themseves as heirs of the 'Bogomils' or, as they see it, 'the Bosnian Church'. That certain motifs and religious symbols were to continue after conversion to Islam may be observed in the *stečci* monuments which have been described by Dr Marian Wenzel in her article 'Four Decorated Steles: the influence of Islam on Bosnian Funerary Monuments', *Journal of Islamic Studies, Islam in the Balkans*, vol. 5, no. 2, July 1994: 221–41. See also Fine and Donia, *Bosnia and Hercegovina*, p. 25: 'Thus, "Bosnian tombstone" is a far more fitting term, for the stones are a distinctive regional feature, produced by Bosnians and Hercegovinians alike, regardless of their Church affiliation. And after the Turkish conquest, many converts to Islam also erected stones which, despite being Turkish/Islamic in shape, continued to display various medieval Bosnian motifs and styles.'

17 Enes Karić, 'In Europe there are no "Indigenous" or "Imported" Religions', paper given at an international colloquium on 'The Encounter between Islam

and Christianity in South Eastern Europe', organized by Dr John Allcock, Cambridge, 1997.

18   Cornelia Sorabji, 'Islam and Bosnia's Muslim Nation', in F. W. Carter and H. T. Norris, *The Changing Shape of the Balkans*, London, 1996, pp. 54–5.

19   Shiism in the Balkans is applied to the Bektashi Sufi brotherhood and to the Babais and Kizilbash in Deli Orman in Bulgaria, and to Alawites in general. The term is occasionally used of members of the Rifaiyya and Khalwatiyya Sufi orders in Kosovo.

20   Edith Durham, *Some Tribal Origins, Laws and Customs of the Balkans*, London, 1928, p. 115.

21   Two classic studies in English, which furnish numerous examples of heterodoxy in Islamic practice in Albania, Turkey and in other Balkan and Levantine regions are John Kingsley Birge, *The Bektashi Order of Dervishes*, London, Luzac, 1937 and 1965; and F. W. Hasluck, *Christianity and Islam under the Sultans*, 2 vols, Oxford, Clarendon Press, 1929.

22   See Florentina Badalanova's chapter in this volume.

23   Enver Hoxha, *Reflections on the Middle East*, London, Workers' Publishing House, 1984, p. 198

24   *Ibid.*, pp. 502–3.

25   Hamid Algar, 'The Hurufi Influence on Bektashism', *Revue des Etudes Islamiques*, vol. 60, 1992, fasc. 1, part 1, p. 45, note 13.

For those who may wish to know more about the course of the spread of these monistic, pantheistic and cabalistic ideas from Persia, Azerbijan, Asia Minor and Thrace, north to Bulgaria and westwards into Albania, see Irène Mélikoff, 'Fazlullah d'Astarabad et l'Essor du Hurufisme en Azerbaydjan, en Anatolie et en Roumélie', *Mélanges Offerts à Louis Bazin par ses disciples, collègues et amis*, edited by Jean-Louis Bacqué-Gramont and Rémy Dor, *Varia Turcica*, Paris, L'Harmattan, 1992, pp. 219–25.

To date, there has been no detailed study of the concepts and poetic images of al-Rumi, Nesimi and other Hurufis, or Iranian literary inspiration on Albanian poets, including those in Kosovo and Macedonia, in any West European language. For a comprehensive study, in Albanian, of their influence on the poet Naim Frashëri see Qazim Qazimi, *Rdikime orientale në veprën letrare të Naim Frashërit*, Prishtinë, 1996, and the great authority, Hasan Kaleshi, 'Vepart Turqisht dhe Persisht të Naim Frashërit', *Gjurmime Albanologijike*, Tiranë, 1968, vol. 1, pp. 143–53.

For those who may wish to know more about the influence of Bektashi art, both popular and classical, in the Balkans, see the extremely well-illustrated publication, F. De Jong, 'The Iconography of Bektashism. A Survey of Themes and Symbolism in Clerical Costume, Liturgical Objects and Pictorial Art', *Manuscripts of the Middle East*, vol. 4, Leiden, Brill, 1989, where there are many examples from Greece, Kosovo, Turkey and the former Albanian-managed *tekke* on Jabal Muqattam, in Cairo.

Part I

Layers of Belief, Blurred Boundaries

1
Prolegomenon to Religion in the Balkans

John Nandriş

Religion represents mankind's highest beliefs and noblest aspirations. Attempts to understand the nature of ultimate reality have motivated many of the greatest achievements of human culture. Religious belief has helped to define such secondary elements of the superstructure as economy and technology, and it has conditioned an impressive variety of social forms. It is religion which makes comprehensible the existence of many of the most important monuments and human achievements which we have inherited from the past, even those which come to us from the depths of prehistory.

South-East Europe *sensu lato*, loosely termed the Balkans, is one of the most varied and ancient core areas of European civilisation. Out of antiquity came belief, followed by codifications of that belief and its full cultural expression, declining in our time into the attempts at its deconstruction which can be seen as a destructive codification of the insecurities of unbelief. Nor would it be right to take a romantic view of the history of religion by ignoring the dark side of belief.

This chapter does not aim to summarise the main religions of the Balkans, but to draw attention briefly to some aspects of belief which emerged from the long perspective of prehistoric antiquity and paganism, and to suggest how in the long term these have profound implications for Europe even today. It does so in part by considering the two cases of a crazy Messiah, and the ballad of a ewe lamb.

One small but remarkable fact is that after all the horrors of the Communist holocaust in Eastern Europe it was possible even during the last two or three years of Ceauşescu's regime in Romania to come across several Christian hermits who had re-established themselves in the remote Carpathians. These monks have by now outlived Communism itself to celebrate the beginning of the third Christian millennium.

The time-depth of religion itself is even longer than this. It has properly to be considered both in terms of its regional variety, and in the uniquely long-term perspective which archaeology takes upon human behaviour. We may recall the words with which Thomas Mann begins his great novel *Joseph and his Brethren*: 'Very deep is the well of the past; should we not call it bottomless?'

Prehistory and identity

The phenomenology of belief penetrates not just back into time, but spatially across Europe. Archaeology can demonstrate how pre-literate man ranged widely across the continent. The Balkans were far less isolated in prehistory than they were under Communism. While the richness of, for example, Romanian ethnography may strike us as something intrinsic, in fact it exemplifies a social and material culture which was formerly much more widespread in Europe. Like so much else in Balkan culture it is fundamentally European in nature.

The Gundestrup cauldron (Fig. 1), with its remarkable pictorial codifications of Iron Age religious beliefs embossed on silver sheeting, was found in a bog in Denmark. It is now felt to have been made in the Balkans during the second or first centuries BC, as part of a long tradition of Thracian relief metalwork (Bergquist and Taylor 1987).

Figure 1.1   The Gundestrup cauldron, c. 100 BC.

The 'Celts' of the La Tène culture of the archaeological record pene-trated into the Balkans from the third or fourth centuries BC, and established a friendly symbiosis with the Iron Age Dacians of Romania. The two peoples were on a rather similar level of not quite semi-literate warrior culture and high craftsmanship. They were prepared to bury their dead in the same cemeteries, which must indicate some reconcili-ation of their respective religious beliefs. Celtic and Dacian religion is complex and lies outside the scope of this chapter.

Because this sort of archaeology lies at the margins of literate classical culture we have literary references to the preoccupation of the Dacians with such topics as astronomy and herbal medecine and the after-life. Some inkling of their religion has thus come down to us, along with the archaeological evidence of their ritual monuments.

At Gradiştea Muncelului in the southern Carpathian mountains there are great rectangular structures, and circles of stone and wood. Most remarkably some of these are a formal replication in andesite of the hut plans of the Dacian *stîne* or sheepfold sites which lie at even higher altitudes (e.g. Meleia and Rudele at 1200–1400 metres). It is fascinating to consider which is emulating the other, and why (Nandriş, 1981).

It has been said that there is a correlation between transhumant shepherds in the Near East and the emergence of monotheism. The life of the shepherd gives him ample opportunity to contemplate the orderly procession of heavenly events, which were in any case perfectly familiar to prehistoric peoples. They integrated their explanatory and religious systems with empirically necessary ones such as the calendars of their agricultural and pastoral cycles. All this is evident from their monuments. It is also clear that the figurine and other ritual material associated with the Balkan Neolithic, like that indeed of Gravettian mammoth-hunters 30 000 years earlier, represents concepts, not merely percepts or children's playthings.

In the earliest Neolithic during the sixth and fifth millennia BC, in the First Temperate Neolithic (FTN), figurines were consistently present but in small numbers. They are regularly associated with items such as clay stamp seals and bone spoons, often in ritual deposits with certain types of painted pottery or artistic motifs such as parallel zig-zags possibly representing water. They take the form of sexually ambiguous figures, susceptible to perception either as females with swelling buttocks and long necks, or as explicitly phallic in themselves. This ambiguity would serve well in a dark hut during an initiation or rite of passage. In later cultures of the fifth and fourth millennia such as Vinča or in the Climax Neolithic cultures of Gumelniţa and Cucuteni, the numbers of figurines are much larger. The intensity of their presence seems to say something to us about a certain weight of religious and social control.

Fifth and fourth millennium Climax societies took the Neolithic mode of behaviour as far as it could go without becoming something else, for example Bronze Age chiefdom or incipient urbanisation. Their settlements could be of impressive size, their houses and artefacts elaborate, and they were often rich in copper and gold. The settlements of Cucuteni or Gumelniţa were of themselves highly demonstrative, almost epideictic, monuments in the landscape, so that we do not find the sorts of megalithic tombs or other earthworks which assume that role in western Europe (Fig. 1.2).

It is clear that gender differences were important as a component of the Neolithic world picture, but it is by no means clear that this involved what literalists like to explain as mother goddesses, symbols of fertility, or the suppression of women. Cultures of such complexity were not without a well-developed social and religious hierarchy. It is simply meaningless ideological conformism to intellectual fashion to call early societies 'egalitarian', or to play down the contributions which religious conviction made to their evolutionary fitness.

Figure 1.2 Reconstruction of a fifth–fourth millennium Cucuteni Neolithic Climax Culture village in Moldavia, drawn by Giovanni Caselli.

Hierarchy in these Late Neolithic Climax societies is indicated by rich burials, and hoards of copper or gold and other objects. An ideology certainly involving dualities of gender is indicated by carefully arranged assemblages of paired male and female figurines, by clay and copper or gold disks and pendants, by house models containing ritual assemblages of furniture and objects, by plastic representations on the ends of houses as in the Cucuteni A2 façade from Truşeşti, by distinctions in the mode of burial between the sexes, and probably in the sophisticated painted pottery designs. The ideology extends to nearly every sphere of life, e.g. in the paired figures which adorn the back of the Cucuteni clay chair model from Lipcani now in Suceava museum.

In a sense the material culture of these people was not 'art' because such an embedded society had no art. Instead everything they did was invested with the significance of their social being. By virtue of our common humanity their culture is enabled to speak, however faintly and distantly, to us. It may be that we should also examine the term 'religion' with more care.

Religion is often invoked in the Balkan situation to explain aspects of ethnic identity and conflict. This can be a dangerous addition to the many literalist misperceptions of the Balkan world. The misery of Bosnia or Kosova is not primarily a 'religious conflict in the Balkans' but a pay-out from the Peace Dividend of Marxism which for long suppressed group identities. The identity of a human group, too, is often seen by essentialists as a sort of polythetic check-list which includes religion. If you are endowed with enough of the characteristics you are presumed to belong to the group. Such and such a religion, dress, language or appearance are taken to define membership.

In contrast to the rather literal 'essentialist' view, scholars such as Fredrik Barth (1969), Elwert (1989: p. 446), or Stephanie Schwandner-Sievers (1998, 1999) favour a more interesting 'formalist' description of ethnicity. This sees it as a formal social action in which the ethnicity is socially constructed. It is a dynamic process in which the criteria of classification may change with time and circumstance, and in which existence and fitness are ratified by usage. This more pluralistic and adaptive approach is reflected in a case-study of group identity among the Jebaliyeh Bedouin of Sinai, and the role of *'behaviour as if'* the member belongs to the group (Nandriş 1990). The antecedents for it may even be sought with care in the work of Jakob Fallmeryer (1835). It is worth remembering the useful linguistic ambiguity of the term *ethnos* in the Greek language, which makes it easy to confuse issues of ethnicity and of nationality.

Moving nearer to the historic present we find expressions of the religious faculty in the Balkans embedded in an ancient social world, which seems to recede ever more rapidly into the past. It was very often harsh, often also humane; but like the characters who inhabited it, it was far from simple. Balkan religion and culture were in part mythological codifications of experience which without being wholly systematic were endowed with important non-utilitarian social functions. The major religions were well aware of ancient beliefs and practices, and often sought to incorporate what they could not entirely repudiate. Nor could they easily eliminate ancient irrationalities. We should look at an example which, as we shall see, links our day to its own irrationalities.

The Messiah

The major players on the Balkan religious scene are familiar enough. Christianity and Judaism established their presence across the millennia, while Byron in 'Childe Harold' reminds us of the role of Islam:

> Land of Albania! Let me bend my eyes
> On thee, thou rugged nurse of savage men!
> The cross descends, thy minarets arise,
> And the pale crescent sparkles in the glen...
> Through many a cypress grove...

The minarets which Byron saw have all but vanished from the landscape, along with the migrating storks which nested on them. His allusion is to the fruitful and momentous interplay between Christianity and Islam in the Balkans over much of the last millennium, especially after the conquest of Constantinople by Fatih Mehmet II in 1453. This has ramifications which are too various to recapitulate, although aspects of them reoccur in later chapters.

The relationships of Balkan history with Judaism are also too complex to summarise here. They do supply one story worth telling, since it has momentous consequences for our long-term perspective on modern pseudo-religions such as Marxism. It is a fascinating chapter in the religion of the Balkans, one upon which such a great scholar of Kabbalah as Gershom Scholem thought fit to invest a surprising amount of energy.

Shabbetai Zevi (1625–76) was a manic-depressive who proclaimed himself to be the Messiah, along with Nathan of Gaza (1644–80) as his John the Baptist, and attracted a huge following. His movement exploded on the scene and stirred up thousands of believing Jews in Eastern Europe largely during the year of crisis 1665–6. This culminated abruptly in a dramatic apostasy when Shabbetai converted to Islam in the presence of the Sultan. His life and apostasy and the movement he led raise grave issues, difficult even now to discuss. However, Scholem has done so and we must also confront them.

It is necessary to know something about the sophisticated Kabbalistic doctrines of Isaac Luria Ashkenazi (1534–72) of Tsefat or Safed in Palestine. Luria was the contemporary of St John of the Cross (1542–91) or of St Teresa of Avila (1516–82).

An earlier seminal work of Kabbalaism, the *Zohar* or flash of lightning, was written in Spain by Moses de Leon (?1250–1300) the contemporary of Meister Eckhart (1260–1328). All owed a great deal to earlier mystics still, e.g. John Scotus Erigena and Dionysus the Areopagite. Kabbalah is not older than but owes much to, and is indeed largely derivative from Christian, Neoplatonist and Gnostic sources, just as the Gospels owe more to Hellenism than to Judaism.

The doctrines of Luria included the idea of *tikkun* or redemption and reintegration of the ideal order. This became necessary following the inextricable admixture of good and evil which resulted from a catastrophic '*breaking of the vessels*' in the course of making several unsuccessful universes (the 'Kings of Edom') during the Creation. We know that Adam notably failed to improve matters, and thus all beings in the world including man came to be clothed in gross physical matter derived from the *qlipoth*, the husks or shells of evil remaining from these early failures, instead of in their original spiritual bodies.

These complex theological situations are perfectly accurately reflected in the painted frescoes of the monasteries of Bucovina, notably at Sucevița. The 'clothing in skins' of Adam and Eve at their expulsion from Eden expresses their assumption of material bodies. The frescoes show the transition from early events in the history of Creation, which take place in a spiritual universe denoted by a white background, to those located after the Fall in the material universe which we still inhabit, for which the background is blue.

This theological accuracy is not wholly surprising. Although remote, the monasteries of Bucovina, to which we shall return below, were in touch with a wide range of philosophical and theological discourse (Nandriș 1970) and were, like Mount Athos, the nursery of many remarkable theologians such as the Abbot of Neamț in Moldavia, Paissy Velichkovski.

Following the amazing neurotic conversion of Shabbetai Zevi, the *qlipoth* of perverted belief itself had psychologically speaking to be salvaged. The degenerate idea of apostasy as an actual means to *tikkun* took hold and spread rapidly, along with all sorts of paradoxical nihilisms. 'We cannot all be saints, let us be sinners.' Some sort of antecedent can be seen, for example, in the apostate *Marraños*, those Jews of the Iberian peninsula who between 1391 and 1498 simulated conversion to Christianity. 'One's heart and mouth may not be one.' The geographically adjacent Cathar *Perfecti* came similarly to feel themselves free from moral constraints, and we should at least be conscious of the possible connections between Balkan gnosticism and Spanish Kabbala, between Cathars and Bogomils (Nandriş, 1992).

In this curious history we see that the perverse thread of revolutionary elitism and deception woven into the ideas incorporated in the French Revolution, and in Russian nihilism, and in the endemic willingness of Lenin and of Marxists in general to lie and to justify means by ends, had extraordinarily early beginnings (Courtois, 1995, 1997). More immediately it gave rise to the sinister radical figure of Jacob Frank (1726–91), who while pretending to Roman Catholicism sought redemption by 'treading on the vesture of shame', a phrase attributed by some Gnostics to Jesus Christ. Frank's preachings and *obiter dicta* constitute 'a veritable religious myth of nihilism' (Scholem, 1974: pp. 315–16: 'Great is a sin committed for its own sake.' The man who might have succeeded Frank as leader of the sect in Offenbach was sent to the guillotine in 1794 under the name of Junius Frey *alias* Moses Dobrushka, *alias* Thomas Edler von Schoenfeld. The subversive idea that he who has sunk to the uttermost depths is the more likely to see the light paved the way for the indecent self-exposure of Jean-Jacques Rousseau, who is so central to left-liberal thought. It even seemed to Wittgenstein (1997) that 'Rousseau has in his nature something Jewish'.

A range of related doctrines became a dangerous vindication of the self-righteous revolutionary ideologies whose echoes ring out through the twentieth century in acts of brutality and terrorism. A penetrating truth about the tiny kernel of such ideas is expressed in Santayana's aphorism, that 'Self-righteousness is the besetting sin of moral indignation'. Such are the multiplier effects of belief.

As Scholem so frankly puts it (1974: p. 320), this anarchical revolt 'was gradually spread by groups...who remained within the walls of the ghetto [and] continued to profess rabbinical Judaism, but who secretly believed themselves to have outgrown it. When the outbreak of the

French Revolution again gave a political aspect to their ideas, no great change was needed for them to become the apostles of an unbounded political apocalypse. The urge towards revolutionising all that existed no longer had to find its expression in desperate theories, like that of the holiness of sin, but assumed an intensely practical aspect in the task of ushering in the new age.'

This was almost the ultimate perversion of the idea of *tikkun*, but not entirely, for Communism and the Soviet Union were still to come. The combination of radicalism and deception (e.g. the deception of Jew by Jew in the ghetto under the pretence of rabbinical orthodoxy) with perverse nihilism (ostensibly motivated by the highest ends), along with acceptance of the validity of subversion of the existing order (combined with high-minded claims for 'liberation'), and mixed with a utopian teleology of final things – all these are important in theological terms for an objective understanding of the emergence of Communism; and of the imposition of Communism in Eastern Europe in which Jews played a significant role; and even of such phenomena as Robert Maxwell and a thousand village Maxwells.

The antinomianism of Sabbatians such as Abraham Perez in Salonika (Scholem 1974: p. 312) paved the way for the Russian no less than for the French Revolutionary. He or she had no idea what to put in the place of the old order but was determined on destruction regardless, in the primitive, superstitious and half-conscious delusion that a Messiah would appear to complete the *tikkun*, and gather up the last few sparks. Evil becomes an integral part of the fallen universe of these ideas. We must, they tell us, all descend like the Messiah into the fallen realm of evil to burst open the doors of the prison from within. Evil must be fought with evil. The 'greedy lust for power' (Scholem 1974: p. 336) which dominated Jacob Frank found equal expression in Russian nihilism, in the epi-semitic pseudo-religion of Marx, and in Lenin's justification of the means by the end.

A redemption of Kabbalistic ideas may be said to have begun during the second half of the eighteenth century, with the emergence of Hasidism and the wanderings of Rabbi Israel Baal Shem in the northern Carpathians. He drew inspiration both from the mysterious and quasi-mythological figure of Rabbi Adam Baal Shem and from the spirituality of the region. Scholem (1974: p. 332 seq.) identifies Adam Baal Shem with Rabbi Hershel Zoref of Vilna and Krakow, author of the *Sefer-ha-Tsoref* and himself a crypto-Sabbatian, who died in 1700 just as the Baal Shem was born. We must now consider a contrasting expression of belief linked to these very regions.

The Miorița

The monasteries of Bucovina, now protected by Unesco, are a creation of Christian Orthodox piety and mysticism, embedded in a region which produced many enrichments of spirituality. The northern Carpathians have enfolded a great many pieties, from the monastic foundations of Stephen the Great (Fig. 1.3) with the hermitages and painted churches of Bucovina to the *shtetl* of the Hasidim. What is potentially of great interest and has been only imperfectly examined heretofore is what may have been the mutual impact upon each other of these spiritualities.

The frescoes are painted outside as well as in the churches, and have survived fierce Moldavian winters. The Last Judgement at Voroneț, reverting finally to the white background which indicates a spiritual environment, shows among the saved in Paradise all categories of men including Islamic Turks and Orthodox Jews. Here Christianity did not condemn whole categories and classes of mankind as did Communism.

During the fifteenth century the walls of the Orthodox monasteries of Bucovina, located among the gentle forested hills of northern Moldavia,

Figure 1.3   Cell of Daniil Sihastru (the Hesychast), Spritual adviser of Stephen the Great (1457–1504), hewn into a boulder at Putna.

were painted in this manner inside and out with glowing frescoes. Plato and other pagan philosophers of the ancient world, are represented in these paintings, alongside the usual Saints and Prophets of Christian iconography (Nandriş 1970).

Pre-Christian beliefs and practices are well embedded in some isolated areas of Romania such as the Maramureş which were Christianised late. They are to be found in the rich segregated traditions of the Romanian shepherd (Fig. 1.4) which Vasile Latiş has brought to life in a profoundly serious work of first-hand ethnographic scholarship (Latiş 1993). This combines the fruits of the controlled scientific imagination with the serious philosophical and sometimes poetic content of the shepherds' own beliefs, in a way which is most necessary for the better humanistic understanding of our topic.

Figure 1.4   Moldavian shepherd in the *cojoc mare* made of five rams skins. In this, one may sleep on snow.

Philosophical paganism takes poetic form in the ballad of *Miorița* or Ewe Lamb. This ballad is quite widely distributed in many variants among Romanian-speaking peoples, and is also found among the Vlahs or Aromâni, the Latin-speakers of the mountains south of the Danube, especially in Greece (Wace and Thompson, 1913). It takes place in a real world, in the sense that it would be possible to illustrate almost any line of it from photographs taken during forty years of ethnoarchaeological work at the sheepfolds of the Carpathians.

The *Miorița* is in short lines of only five or six syllables. Such material is notoriously difficult to render convincingly in translation, but freely speaking:

Pe-un picior de plai,	From the pastures' feet
pe o gură de Rai,[1]	In a landscape sweet
iata vin în cale[2]	See, down trackways wending
și cobor la vale,	To the vale descending
trei turme de miei	Flow three flocks of sheep[3]
cu trei ciobanei.	Whom three shepherds keep.
Unu-i Moldovan	One is Moldavian
unu-i Ungurean[4]	one is Transylvanian
și unu-i Vrâncean...	and one from Vrancea land...

(© J. G. Nandriș)

1 Literally at heaven's mouth.
2 Latin *callis*/Italian *calle* = *tratturi*, broad swathes of the drove roads for transhumant sheep.
3 Literally, the lambs. Flocks are divided with milking ewes, non-milkers (*șterpi*), and rams all pastured separately.
4 From Ungur country; not with the connotation of Hungarian.

The opening lines describe the meeting of three shepherds with their flocks on the flanks of the mountain pastures. They come from three of what are still major centres of pastoralism. The shepherd from Moldavia (Fig. 1.4) has a curly-haired pet ewe-lamb, the *Miorița* (diminutive of *Mioara* = ewe). The two other shepherds, from Vrancea and Transylvania, envy the Moldavian his rich flocks of curly sheep, his spirited horses, and his proud dogs. They plot to murder him at sundown. The lamb overhears them conspiring. She is put off her food and comes bleating to her master. She begs him to draw the flock out to some strategic place, and summon his fiercest dogs to defend him.

The reaction of the Moldavian shepherd is perhaps surprising. He instructs the lamb that if in reality he is to be murdered that night in the millet field, then they should bury him close to his sheep pens, behind the *stîna*, so that he shall be near his flock, and within hearing of his dogs. (Millet was a staple before the introduction of maize from the New World.)

> His flutes are to be placed beside him....
> a small pipe of beech
> with its soft sweet speech,
> a little flute of bone
> with a loving tone,
> and one of elderwood
> fiery-tongued and good
> .... where the wind may play in them and the
> sheep can come to listen (Fig. 1.5).

Figure 1.5   Piping shepherd depicted on a Romanian shepherd's staff of Cornelian cherry.

The lamb is not to reveal that he has been so brutally murdered, neither to the sheep themselves, nor to his old mother whom he knows will come looking for him. It must say rather that he has gone away to marry a princess; that at his wedding the high mountains were his priests, the pine trees and maples his witnesses, that the stars came out as torches and the other heavenly bodies played their part, while the birds sang for him as musicians.

The reaction of the Moldavian is fully European in tone. This ballad does not sustain the primitive tribal morality of 'an eye for an eye' which so fatally detracts from the achievements of early monotheism in the Near East. While the other shepherds are envious, envy is not institutionalised as it was in Communism. Nor does the *Miorița* mirror the brutal *machismo* of the Mediterranean world, which finds its political expression in revenge, nationalistic violence, bombastic male honour, and the paranoia which piles blame on others. The resignation to death which is expressed here may to some extent incorporate overtones of the Pythagorean philosophy which is supposed greatly to have influenced the Dacians during the Iron Age in Romania; and it mirrors the proverbial resignation of the Aromâni south of the Danube in the face of social victimisation.

The *Miorița* just as profoundly expresses the contemplative nature of the Moldavian who is the central figure. His supposed fatalism derives from centuries of survival techniques. He lives at a crossroads of Europe subject to centennial invasion and pillage by Eurasiatic barbarians of which the Russian invasion of Bucovina after the Molotov–Ribbentrop pact in 1939–40 was merely the most recent expression.

In response to the argument that fatalism is a poor political philosophy, it has to be said that this culture has survived centennial invasions, and more intensive attempts at its destruction than, for example, the islands of Britain over a comparable period. It seems therefore that quite apart from their ethical inferiority neither *machismo* nor revenge (an eye for an eye) can be wholly necessary or effective in evolutionary terms.

To represent this philosophy as fatalist is indeed facile, and is typical of the superficiality of literalism. On the contrary the ballad can be seen as a display of self-possession, an assertion of strength. The response of the Moldavian shepherd incorporates a studied overcoming of death by casting out fear. It is a contrast of what is noble with what is base. The role of injustice is to teach us not to look back; not to do evil in return for evil. The contrast with the set of ideas examined previously is stark indeed.

The *Miorița* looks back to that fervent belief in the afterlife which Herodotus records as one of the preoccupations of the Dacian peoples of this area. Characteristically for Romanian peasant belief, it draws on the phenomena of nature and the close relationship between the shepherd and his flock in order to question the relationship of man to creation.

It could be seen from the *Miorița* alone that beneath later developments in the history of Balkan religion there lies a deep stratigraphy, at first pagan and in contact with literate classical civilisation, and deeper down with pre-literate but not ineloquent prehistoric cultures. Ideally we ought also to deal more fully than is possible here with the religions of classical antiquity. Both Greece and Rome profoundly influenced the Balkans and, as with the story of Orpheus, this influence was surely reciprocal.

A deep stratum of beliefs which could be characterised as superstitious rather than religious is still extant in Romanian popular culture, much of it deriving from the Roman occupation of Dacia. If this seems unlikely let us remember that Peter and Iona Opie have demonstrated the descent of some children's games from Roman times quite independently of adult mediation (Opie and Opie, 1969). Other still extant phenomena such as fire-walking* are pagan, while the masked dancers of the Balkans and indeed Europe are also clearly of prehistoric antiquity (Seiterle, 1984, 1988; Coulentianou, 1977; Sartori 1983).

Conclusion

Clearly an overview such as this cannot do justice to such important topics. It would have been presumptuous to try and recapitulate major religions and systems of belief, and this is not what we set out to do. Balkan religions, like the rich cultures of its peoples, have consistently exhibited an ability to reconcile disparate elements in a harmonious manner whether through syncretism or symbiosis. Too often it has been politicians seeking power, and ideological purists seeking illusory consistency and exclusive solutions, who have created conflict.

The evolving temporal dimension of belief finds an evocative expression in the great underground cisterns of Byzantium, built for the Emperor Justinian by seven thousand slaves in the sixth century to

---

* Firewalking still takes place at Langadha in Greek Thrace and at Agia Eleni near Serres at the festival of Anastenaria on 21 May, the feast of Constantine and Helen. Until recently firewalking used to occur in Bulgaria at some villages in the Strandzha Mountains, on 3 June. These two dates are thirteen days apart, which is the difference between the old Julian and the new Gregorian calendars. In other words, these rituals related originally to the same date and hence to each other. Virgil refers to the rite in the context of devotion to Apollo in *Aeneid* XI: 1074–5 (trans. Robert Fitzgerald).

supply water to the city. They are covered over with arches of brick, supported by the 'thousand and one' marble columns figured in the name of the cistern of *Bin bir direg*. Many of these were brought from the Hellenistic ruins of Asia Minor.

The cistern of *Yerebatan Saray* is known as the *Underground Palace*. A forest of 336 half-submerged columns nine metres high recedes in serried ranks 140 metres into darkness which echoes to the slow drip of water from the roof. Against the rear wall of the cistern there lies a hidden reminiscence of antiquity, the expression of many syncretisms linking Byzantium with the religions of antiquity. A wavering torch beam reveals under water a number of powerfully carved heads of stone Medusae, each more than a metre high. These were brought from Didyma in Ionia to be re-used in 542 AD as column bases in the furthermost reaches of the cistern (Fig. 1.6).

Figure 1.6   Medusa from Yerebatan Saray cistern, Byzantium, showing its re-use as an underwater column base.

The devout citizens of Byzantium were drinking from dark waters, in which the half-remembered achievements of paganism lay submerged, as if to symbolise the hidden accretions of belief and the solemn passage of time which have structured Balkan religion.

References

Barth, F. (ed.) (1969) *Ethnic Groups and Boundaries* (London: Allen & Unwin).

Bergquist, A. and Taylor, T. (1987) 'The Origin of the Gundestrup Cauldron', *Antiquity*, 61: 10–24.

Coulentianou, J. (1977) *The Goat Dance of Skyros* (Athens: Ekd. Ermis).

Courtois, S. (1995) *Histoire du parti communiste français* (Paris: Presses Universitaires de France).

Courtois, S. (1997) *Le livre noir du communisme* (Paris: Pierre Laffont).

Davies, N. (1966) *Europe* (Oxford: Oxford University Press).

Elwert, G. (1989) 'Nationalismus und Ethnizität: Über die Bildung von Wir-Gruppen', *Kölner Zeitschrift für Soziologie und Sozialpsychologie*, 42(3): 440–64.

Fallmeryer, J. (1835) [Pt II 1830] *Geschichte der Halbinsel Morea während desMittelalters: Lehre über die Entstehung der heutigen Griechen* (Stuttgart and Tübingen).

Latiş, V. (1993) *Pâstoritul în Munţii Maramureşului: Spaţiu şi Timp* (Baia Mare: Marco & Condor).

Nandriş, G. (1970) *Christian Humanism in the Neo-Byzantine Mural Painting of Eastern Europe* (Wiesbaden: Otto Harrassowitz).

Nandriş J. G. (1981) 'Aspects of Dacian Economy and Highland Zone Exploitation', *Dacia* (Bucharest), 25: 231–54.

Nandriş, J. G. (1990) 'The Jebeliyeh of Mount Sinai and the Land of Vlah', *Quaderni di Studi Arabi*, 8: 45–80, Figs 1–16 (Universitá degli Studi di Venezia, Dipartimento di Scienze Storico-Archeologiche e Orientalistiche). Casa Editrice Armena.

Nandriş J. G. (1992) 'Aromâni and Bogomobility: Some Thoughts on *stečci*, Aromâni, Bogomils, and the Transmission of Ideas through Pastoral Mobility', in Gabriele Birken-Silverman, Thomas Kotschi and Gerda Rössler (eds), *Beiträge zur sprachlichen, literarischen und culturellen Vielfalt in den Philologien*: 15–71. Festschrift für Rupprecht Rohr zum 70. Geburtstag, Stuttgart, 1992.

Opie, Iona and Opie, Peter (1969) *Children's Games in Street and Playground* (Oxford: Clarendon Press).

Sartori, D. and Lanata, B. (1983) 'Arte e tecnica della maschera in cuoio', *Arte delle Maschera nella Commedia dell'Arte*, 1983/84: 192 et seq.

Scholem, G. G. (1974) [1941, 8th printing] *Major Trends in Jewish Mysticism* (New York: Schocken Books).

Scholem, G. G. (1987) *Origins of the Kabbalah*, ed. Werblowsky (Princeton: Princeton University Press). Update and translation of *Ursprunge und Anfänge der Kabbala* (1962).

Schwandner-Sievers, S. (1998) 'Ethnicity in Transition: the Albanian Aromanians' Identity Politics', *Etnologica Balkanica* 2: 167–84.

Schwandner-Sievers, S. (1999) 'The Albanian Aromanians' Awakening: Identity Politics and Conflicts in Post-Communist Albania', *European Centre for Minority Issues, Working Paper*, no. 3.

Seiterle, Gérard (1984) 'Zum Ursprung der griechischen Maske, der Tragödie und der Satyrn. Bericht über den Rekonstruktionsversuch der vortheatrale Maske', *Antike Kunst* (Basel) 27 Jhrg., Heft 2: 135–45 (& Plates 15–19).

Seiterle, G. (1988) 'Maske, Ziegenbocke und Satyr', *Antike Welt* (Basel) 19 Jhrg.: 2–14.

Wace, A. J. B. and Thompson, M. S. (1913) *The Nomads of the Balkans* (London: Methuen).

Wittgenstein, L. (1977) *Vermischte Bemerkungen* (Frankfurt-am-Main: Suhrkampf).

2

Interpreting the Bible and the Koran in the Bulgarian Oral Tradition: the Saga of Abraham in Performance

Florentina Badalanova

Frame of reference: oral exegesis of the sacred scriptures

An understanding of the correlation between 'popular beliefs' and the religions of the Book – the Bible and/or the Koran – is an important component in a coherent picture of cultural and political processes in South-Eastern Europe today: the oral tradition may be seen both as an implicit bridge between different confessional patterns and as a verbal paradigm of popular faith.[1]

In the Balkans, folk religion is predominantly expressed in the Christian and Islamic hypostases. While they diverge radically at the explicit level, at the implicit level they often converge, forming a phenomenon one could define as a confessional stereotype in the making.

My approach to this question is anthropological.[2] I focus on the oral exegesis of the Holy Scriptures, with special regard to the saga of Abraham, as performed today amongst Christians and Muslims in the Balkans.[3] I explore a corpus of folklore materials, recorded over the last century and a half, which either refer to the biblical account of the filial sacrifice, or relate to its Koranic counterpart, thus illustrating the vernacular paradigm of the encounter between Christianity and Islam. Analysis of their content provides evidence that in the Balkans, apart from the explicit interaction between the two basic confessional paradigms, there was also an implicit interplay between Jewish-Christian and Jewish-Muslim encounters. It is particularly significant that some of the folklore interpretations of the saga of Abraham contain narrative components which are to be found neither in the Bible, nor in the Koran, but in 'Haggadah' and 'midrash' (rabbinic legends and homiletic writings). In fact, among the Eastern Orthodox Slavs in the Balkans there is a corpus of traditional folklore texts concerned with the theme of Abraham's sacrifice that refer more obviously to some midrashiim and to Mohammed's account, than to the biblical pattern. At the same time, some folklore versions of the saga of Abraham, as performed in a Muslim environment, are much closer to the biblical or midrashic tradition than to the established Koranic text itself. In a sense, these materials can be looked upon as 'living antiquities', reconciling the three Abrahamic religions; in other words, they do not merely represent their theological divergence, but also exemplify their common origin, thus offering heuristic answers to the questions raised by scholars searching for 'an interpretation both faithful to the text (of the Sacred Scriptures) and to the community of believers'.[4] In this way the empirical folklore facts can become factors elucidating significant aspects of pre-literary tradition that preceded the written appearance of both the Bible and the Koran. On the other hand, the results of the field-research amongst rural communities in the Balkans indicate that the oral tradition, preserved by illiterate story-tellers – whether of Christian or Muslim origin – represents an intriguing cultural phenomenon, in which the three religions of the children of Abraham are often interwoven.

The descriptive analysis of these folklore accounts can play a crucial role in further anthropological research into the oral dimensions of the religions of the Book. Indeed, I believe it should be the starting-point for any critical evaluation of the vernacular scope of the Holy Scriptures. It can also explore the various transformations in the content of the biblical or Koranic saga of Abraham, and help explain these alterations in the context of comparative religious studies. Moreover, analysis of the folklore materials at the syntagmatic level can elucidate the way the singer of tales combines patterned structures from different cultural and confessional paradigms, forming one linear text which is to be performed in accordance with the canon of the local tradition. This analysis may also illuminate a forgotten stage of the traditional oral development of the biblical and Koranic saga of Abraham. One may suppose that it was at that stage that a number of motifs began to gather round the name of the protagonist, thus forming the plot of the saga of Abraham, as later presented in the Holy Scriptures. This is an example of the way the oral hypostases of the sacred books came into being. On the other hand, it shows how the Christian and Islamic folklore cycles of songs and narratives about the life of Abraham coexist in any local tradition. In other words, the examination of these folklore materials at the syntagmatic level can elucidate the correlation between the

Judaic, Christian and Muslim traditions at the paradigmatic level. In fact, the study of the 'morphology' of the saga of Abraham may also reveal the specific underlying rules of the 'transformational grammar' that generated the basic cultural and confessional patterns of the three monotheistic religions. In other words, these rules had shaped a certain common initial paradigm from which Judaism, Christianity and Islam eventually sprang.

Slavia orthodoxa recounting the book of Genesis

Slavonic folklore versions of the biblical saga of Abraham, contained in Genesis (11:26–25:18), demonstrate the correlation and interdependence between the medieval written tradition – both canonical and apocryphal – and ethnopoetics. According to the paradigm of Christian allegory, Abraham is regarded as a model of Christian faith: his readiness to sacrifice Isaac is understood as a symbol of the sacrifice of God the Son, Jesus Christ, by God the Father.

For Bulgarian folklore culture, the songs of Abraham's sacrifice provide the basis of the setting for the 'Kurban' (ritual sacrifice), and this custom has strict time-related dimensions: it is considered to be on the temporal threshold of the 'Merry Green feasts'. In most villages it is solemnised either on St George's Feast Day (23 April or 6 May), or at Easter. In some regions, however, it may be performed at Pentecost, thus implicitly referring not only to the Christian festival that takes place on the seventh Sunday after Easter, but also echoing the cult settings of Jewish celebration of the harvest, performed fifty days after Passover. Similarly, in some Christian villages in the Valley of the Roses (in the vicinity of the city of Kazanluk, Central Bulgaria), the song of Abraham's sacrifice is sung by women reaping the harvest, while in some regions of south-eastern Bulgaria it is performed on the feast day of the Prophet Elijah (20 July or 3 August), thus marking the end of the harvest. In fact, almost everywhere in Bulgaria, Christians mark the story of Abraham, with nearly every village having its own feast to celebrate the Great Sacrifice. On this day, the oldest man of each family in the village where the celebration is taking place presents a sacrificial offering to God, thus re-establishing the bond between his home and the household of the biblical patriarch. Mythopoeic imagery transforms narrative biblical patterns into a ritual performance, while the actual performers involved in the mystery acquire, during the sacred ceremony, the status of the biblical characters. Any male participant in the 'Kurban' feast repeats the gestures of the biblical patriarch while slaying the lamb, thus symbolically partaking in the Old Testament drama. In fact, any man performing the sacred ritual of 'Kurban' sacrifice does not merely act 'like' Abraham, he 'becomes' Abraham and acquires his status as a patriarch of the clan. Through the power of ritual, he is transformed into him by faith, and, by faith, 'becomes' the 'father of multitudes' (Gen. 17:15). Lastly, by virtue of his obedience, he is given God's love and benediction: 'That in blessing I will bless thee, and in multiplying I will multiply thy seed as the stars of the heaven, and as the sand which is upon the sea shore; and thy seed shall possess the gate of his enemies; And in thy seed shall all the nations of the earth be blessed; because thou hast obeyed my voice' (Gen. 22:17–18). On the other hand, the biblical formula 'as the stars of the heaven' has its equivalent in the formulaic language of the Slavonic and Balkan oral tradition.[5] This verbal expression functions as one of the highly conventional proems or endings framing classical Christmas carols, and especially those sung to the oldest male member of the family, to the 'Lord of the House', or his offspring:

Тебе пеем, Добæр јунак,	We sing for you, good hero,
Тебе пеем, Бога славим:	We sing for you, glorifying the Lord:
Колку звездина небету,	May this household be brimful with health
Толкуз здрави в тази кæщта.[6]	As the heavens are brimful with stars!

Here, 'the paterfamilias' is panegyrised as an earthly counterpart of 'Our Father in Heaven', yet, as this text emphasises, the actual 'addressee' of the Christmas carol is the Lord Himself. Sung during the night in which His Son is born as 'a man amongst men', the biblical expression 'as the stars of the heaven' manifests not only the encounter between the terrestrial and celestial realms, while Christ's deified humanity descends 'now' amongst 'us', but also signifies the forthcoming drama of filial sacrifice, and indeed the 'sacrament of sacraments' – the Eucharistic mystery.

It is appropriate to note here that in the biblical saga of the filial sacrifice, as recounted in the book of Genesis, there exists another reference to the 'star' topic. It precedes the episode concerned with the birth of Isaac. Thus, when 'the word of God came unto Abraham in a vision, saying, Fear not, Abraham: I am thy shield', the prophet is told: 'Look now toward heaven, and tell the stars, if thou be able to number them: and he said unto him, So shall thy seed be' (Gen. 15:5).

It is particularly significant in this respect that in some medieval extra-biblical sources Abraham is also called 'the one who tells the stars', which implicitly refers to the image of an oracle or soothsayer who 'knows' the divine language of Heaven, and who can, thus, 'talk' to the sky, and to whom God listens and replies.[7] This detail suggests that the character of Abraham, as represented in the book of Genesis, might have inherited some archaic functions of the prophetic figure, as conceptualised by the pre-literary Semitic tradition. On the other hand, as the Church Slavonic texts show, a similar tendency may be seen in some apocryphal versions of the saga of Abraham in which 'the star topic' serves as a proem of the narrative. Thus, according to a text entitled 'Sermon on Righteous Abraham' (Slovo o Pravednago Avraama), while contemplating the divine act of creation of the Universe as a manifestation of God's nature, the prophet (the son of a certain idol-maker) 'looked toward heaven and told the stars'.[8]

The story of how the Prophet Abraham, the son of an idol-maker, contemplated the mysteries of the Universe and rejected the religion of his earthly father, is also considered to be one of the most widespread legends narrated amongst the Muslim and Jewish communities. Thus, according to J. E. Hanauer, who visited the villages of the Holy Land collecting yarns spun by the local people, 'these are tales told by both Jews and Moslems'.[9] Besides, in the Koranic text, there is also a direct indication that, after having said to his father Azar, 'Takest thou idols for gods? I see thee, and thy people in manifest error' (Koran, Sura 6, vv 73–4), Abraham was shown by Allah 'the kingdom of the heavens and earth, that he might be of those having sure faith' (Koran, Sura 6, v 75). In fact, according to classical Islamic tradition, the divine revelation took place 'when night outspread over him', and the prophet 'saw a star and said, "This is my Lord"' (Koran, Sura 6, vv 75–80). There is clearly a striking parallel between the Slavonic apocryphal tradition and the Koranic text. Thus, according to Mohammed's account, after having observed the Heavenly bodies rising and setting, Abraham utters: 'I have turned my face to Him who originated the heavens and the earth, a man of pure faith; I am not of the idolaters' (Koran, Sura 6, v 79), whereas the Slavonic apocrypha give a concise version of the episode concerned. It is particularly important, however, that 'the star topic' serves as a framework of the plot in both traditions. Furthermore, the dialogue between the Prophet and the stars has been traditionally included also in the Bulgarian folklore interpretations of the saga of filial sacrifice. This formulaic expression usually functions as an introduction to the plot of a series of texts,[10] portraying Abraham's future as a mysterious riddle 'in the stars', thus resembling not only the traditional folklore conceit of fatherhood, as represented in Christmas carols, but also the implicit – or perhaps one should say 'coded' – biblical and Koranic designation of the 'Father of multitudes', as progenitor of offspring who would be as numerous as the stars of Heaven. In general, these ritual songs start with a formulaic dialogue between the Morning Star (zvezda zornica) and the praying parents who promise the star that if God gives them a son, his name will be 'the One who Belongs to God' (Bogumho) and he will be sacrificed to Him.[11]

On the other hand, according to popular Christian tradition, the ritual sacrifice of the lamb as a divine substitute for Isaac, is regarded as a functional counterpart of the Eucharist. Isaac resembles Christ, the Lamb of God, who has voluntarily sacrificed himself for the forgiveness of the sins of men. It is indicative that in Slavonic languages there exists an unequivocal similarity between the word denoting 'lamb' (Bulg. 'agne', Serbo-Croat 'jagnje', etc.), and the liturgical formula 'Lamb of God' ('Агнец').[12] This similarity is significant from the point of view of oral tradition. It illuminates the revelation that this is He – the AGNUS DEI ('Агнец ьт') who is indeed the sacrificial lamb ('жертвеното агне') – and vice versa. In other words, God the Son is to be sacrificed by God the Father through the image of a lamb, and as such He will touch the realm of men. His blood, the blood of the 'Lamb of God', will descend to earth, flowing out of the body of the slaughtered lamb, redeeming those who partake in the mystery of His sacrifice. Thus the folklore exegesis reveals – in terms of ethnopoetics – the mystery of the Eucharist, connecting it with the ritual of the 'Kurban' feast. Evidently, the ritual of the Great Sacrifice, as performed amongst the Christian communities in the Balkans, is regarded as a sacred undertaking embedded in the biblical paradigm of righteous behaviour, as established by Abraham.

As far as the Christian songs interpreting the mytheme of Abraham's sacrifice are concerned, one the earliest versions was recorded and published by the Miladinov brothers in 1861.[13] Two years later P. Bessonov published another account,[14] which was followed by a series of texts registered by V. Kachanovskii,[15] N. Bonchev,[16] etc. These variants, together with the texts recorded and published over the last century and a half,[17] indicate that the 'classical' plot of the saga of Abraham, as performed amongst Christians in South-Eastern Europe, consists of the following episodes:

1. The prayer of Abraham the Hospitable to God
2. The three visitors at Mamre

3. The Blessing: 'You will be given a child!'
4. The birth of Isaac
5. The sacrifice of Isaac
6. The Divine Intervention
7. The Origin of the 'Kurban' custom as a Christian ritual.[18]

As regards the age of the son, the folklore versions of the saga of Abraham suggest that when the time for the Great Trial had come, he was on the threshold of his boyhood and not an infant anymore. It is significant, however, that according to these texts the boy is to be sacrificed when he reaches the age of being able to collect wood. The canonical account of Abraham's sacrifice in the book of Genesis does not contain any particular information regarding the age of 'his most beloved' son, whereas the folklore versions of the biblical saga of the filial sacrifice put a particular emphasis on this topic.

Thus, according to the text of the song I recorded in 1977 in the village of Glavan, Southern Bulgaria (Thracian lowlands), the childless Abraham (Avram) prays to the Lord:

> 'Give me, My Lord, give me
> An offspring from my heart,
> To walk around the courtyard,
> To say "Mother!", and "Father!",
> To go then to the field,
> To go to the field and plough it,
> And to fetch a cart full of firewood . . .
> I vow to slaughter him then as a sacrifice
> To the Lord God and to Saint George!'
> God stood there listening,
> And they had an offspring from the heart,
> And they gave him a Christian name,
> a Christian name, after the name of Saint George,
> And the little boy named Georgi grew up
> and became a fifteen years-old youngster.
> And they sent him to the field,
> To the field, to plough it,
> To fetch a cart full of firewood,
> Of firewood and of flour.
> When he came back home,
> His mother was baking bread,
> Baking bread and weeping.

> His father was whetting knives,
> whetting and weeping.[19]

In other words, according to the text of this ritual song, the boy is intended to be sacrificed not only when he reaches the age of being able to plough the land of his father, i.e. to perform classical male agricultural activities, but also when he is able to collect wood. There is a similar approach to the age topic in Islamic oral tradition, as preserved amongst the Muslim people in Bulgaria, and its framework recalls the substructure of the classical Koranic account. Thus, after his having been given 'the good tidings of a prudent boy', the Prophet is asked to sacrifice his offspring when he reaches 'the age of running with him' (Koran, Sura 37, vv 99–101). In the folklore interpretations of the Muhammad's account, however, the age topic appears to be closely related to the wood topic, and this is spelled out not only explicitly, but also implicitly. Thus, according to some folklore narratives, Abraham intends to sacrifice his son in the forest, after the boy has collected wood for the burnt offering there. The content analysis of these texts indicates that what is regarded as the most important component of the plot is not that the father and the son go to collect wood, but the fact that they go to the forest. This establishes a straightforward framework portraying the forest as an emblematic setting for the filial sacrifice. On the other hand, this fragment recalls the folklore concept of the forest as a classical locus for performing the mystery of initiation. Thus, via the reference to 'the wood' topic, the forest appears to be recognised by the storytellers – because it is mythologically transparent – as not only an obvious setting, but as the ultimate scene for the drama of the Great Trial. The above-quoted song from the village of Glavan,[20] together with the Muslim folklore narratives, as recorded among rural communities in Bulgaria,[21] appears, then, to be a descendant of a version of Abraham's saga that preceded the written appearance of the Holy Scriptures. It was probably transmitted only orally amongst the People of the Book. It is most significant that the detail regarding the age of the son who was intended to be sacrificed by his father, was included neither in the Koran, nor in the Bible, yet it has survived in the folklore tradition of Muslim and Christian people in the Balkans as an enduring memory of the oral hypostases of the Holy Scriptures. In fact, the fragment related to the 'wood' mytheme seems to bear certain encoded traces of some hidden stages of the diachronic development of the saga of Abraham, thus exemplifying, yet again, the encounter between Christianity and Islam at the popular level. On the other hand, these folklore texts depict the interaction between the oral hypostases of the religions of the book, thus casting light on their common roots.

It is evidently not only the oral tradition, whether Christian or Muslim, that considers 'the wood' topic to be of a particular importance for the internal logic of the saga of filial sacrifice. Early patristic tradition also acknowledges it. Christological exegesis interprets the figure of Abraham's son who carries the wood on his shoulders while climbing the hill where he is to be sacrificed as a foretype of Jesus carrying his cross towards Calvary.[22] This typology was extensively developed by the Church Fathers, and by Irenaeus, Tertullian and Origen in particular.[23] Hence, in Christian convention the figure of Isaac carrying wood on his shoulders has traditionally been depicted in connection with the Crucifixion. Isaac is, then, regarded as a foretype of Jesus Christ, while Abraham is the earthly counterpart of God the Father. The way in which popular faith spells out this typology is quite specific, but it is nonetheless evident that the folklore renderings of either the biblical or the Koranic versions of the saga of Abraham consider this mytheme to be of great importance.

On the other hand, analysis of the texts of South Slavonic songs accompanying the 'Kurban' custom shows that they do not always associate the motif of the sacrifice of a son by his own father with the actual names of the biblical patriarch Abraham and his offspring Isaac; but the functional parameters of these songs are strictly regulated. The relationship between their performance and the ritual sacrifice of the lamb, designed for the common feast, is obligatory. Accompanying the act guaranteeing the life, fertility and good health of each clan of the village and of the village as a whole, these songs are mostly performed at the feasting table or to a 'Horo' chain-dance, thus unfolding the folklore aetiology of the ritual acts related to the sacrificial offering. It should be emphasised that before or after singing the tale of filial sacrifice, the informants usually recount the biblical legend of the Great Trial as well, thus commemorating 'the godly prophet and his offspring'. According to these extended oral narratives, 'all the old Christians before Abraham' used to offer their children as a sacrifice to God, and by His remarkable intervention, the Lord ordered that a lamb and not a child be presented to him. The ritual reminder of the slaying of boys as a sacrifice to God has been updated via the symbolic 'blood-staining' of children's foreheads with the sign of the cross made with warm lamb's blood. The blood from the sacrificial lamb will guarantee them life and good health. Here it is appropriate to note also that the ritual of offering a lamb, as performed amongst the rural communities in the Balkans, contains some allusions not only to the saga of Abraham (as presented in the Old Testament) but, apparently, also to some themes in Exodus 12. According to the biblical text, before the Children of Israel escaped from Egypt, they were commanded to offer a lamb. In a similar way, the storytellers emphasise that 'in the days of yore', before the time of the 'the godly prophet Abraham and his offspring', people used to slay their children as a sacrificial offering to God. Correspondingly, the drama of the Great Trial is portrayed as a sacred precedent interrupting the pattern of the old law, thus establishing a new creed. Moreover, the songs recounting its divine constitution function as an oral exegesis of biblical events, thus serving as living memory of 'the time of the sacred Metamorphosis'.

Slavia orthodoxa recounting 'Midrashim' and the Koran

Along with the stories describing the aborted slaying of Isaac, as 'God wants no human oblation', there are also some texts in South-Slavonic Christian folklore, predominantly songs, in which Abraham does actually slay his son as a sacrificial offering to God.[24] These accounts correspond to some 'midrashim' in which the theme of the accomplished filial sacrifice serves as a framework for the plot.

As Seth Daniel Kunin points out, there are numerous midrashic collections that include various texts interpreting the motif of the death of Isaac. In his analysis of their content, he stresses that this episode 'is developed in several ways in differing texts'.[25] He also mentions that, according to the account of 'Midrash HaGadol', which is considered to be one of the most popular thirteenth-century collections: 'the death and resurrection is emphasised by a passage stating that Isaac had spent three years in Paradise before returning to his home. The midrashic tradition also directly ties Abraham's returning alone to the death of Sarah: Sarah dies when she sees that he arrives without Isaac.'[26]

According to some scholars (Kunin, Spiegel), this detail had considerable impact on the further evolution of the midrashic tradition. Similarly, in the text of 'Lekah Tov' (a thirteenth-century 'midrash'), it is stated that 'Isaac died and was revived by dew drops of resurrection', while the 'Shibbole HaLeket' midrash (another collection from the same period) asserts that 'Isaac was reduced to dust and ashes, after which he was revived by God who used life-giving dew'.[27]

These midrashic collections are indicative of a particular tendency in the medieval Judaic tradition; according to Kunin again, 'in these later texts the trend begun with the Bible ... finds its logical conclusion. The structure develops from a suggestion of sacrifice in the biblical text, to a literal sacrifice and resurrection in the midrashic texts.' In the Balkan oral tradition, however, Isaac is resurrected after having been placed in the oven. The motif of 'the wondrous child in the burning oven' who is miraculously resurrected in the flames, thus returning from death to a new, 'eternal' life, in order to proclaim a new law and a new convention, serves as their focal point. Besides, the thematic range of folklore texts concerned with the story of the ritual slaughter of the child, and with the episode about his resurrection from the live embers of the oven in particular, also corresponds to some texts from the late Judaic tradition. In these, the new-born prophet becomes a victim of persecution on behalf of King Nimrod, frightened by astrological predictions. The King orders his people to kill all new-born children, and Abraham miraculously survives; subsequently, when the king insists that Abraham bow both to the idols and to him, and the prophet refuses, Nimrod throws him into a burning oven, from which the Lord rescues him.

As Joseph Gutmann points out, 'this story appears in Hebrew manuscripts of the fourteenth century and is also found in coeval Christian art, the *Bible Moralisée* and the *Speculum Himanae Salvationis* manuscripts'[28] (Gutmann 1979: 11).

It should be noted that the motif of 'the wondrous child in the burning oven' who is miraculously resurrected in the flames, also correlates with the Koranic text about Abraham cast into the flames (Sura 21, 'The Prophets', vv 64–70; Sura 29, 'The Spider', vv 20–5 and Sura 37, 'The Ranks', vv 91–100), and it is in this light that the significance of folklore tradition can be appreciated. It is indicative that there exists an analogy between the texts of the late Judaic tradition and the Koran, on the one hand, and the Bulgarian folklore texts[29] on the other, and these correspondences are intriguing. They show that the Slavonic and Balkan oral tradition still remembers the prototexts that had generated not only the Bible and the Koran, but also some midrashic texts. These folklore renderings show, however they are recounted, that the saga of Abraham functions as the backbone of the ritual life of the 'children of the Book', and as a framework of their sacred genealogy.

As far as the Christian communities in the Balkans are concerned, the life of Abraham is shared by them each and every year, thus becoming a focal point of their most important ritual celebrations. The sacrifice of Isaac seems to be happening and to be re-experienced each time, bringing to mind the Christian commitment to the biblical event and to the destiny of the eleventh Patriarch after the flood. Abraham becomes a 'relative', and, of course, 'ours' by nationality. Indicative in this respect is the substitution in some songs of the name of the biblical patriarch by Slavonic names: Stoyan, Lazar, Ivan, etc. In this way the biblical narrative is transformed in folk-memory. At the same time, every village, observing the sacrifice of the son of Abraham, turns into Canaan. Genesis is built into the real life of the village community and Old Testament legends become folkloric aetiological text.

The Koranic Saga of Ibrahim re-narrated

The folklore narratives concerning the ritual of the Great Sacrifice recorded among Muslims in the Balkans function, as a rule, in the same way as their Christian counterparts: as oral hypostases of the Holy Scriptures. It is appropriate to note here that the term 'Koran' refers to the concept of 'recitation' (*Qira'a/Qur'an*), while Allah is considered to be 'The Speaker'.

In Muslim communities in Bulgaria, the prophet who intends to sacrifice his son is known as Ibrahim (or Ibraim) Peigamber, Ibriam, and even Ismail. In some tales, however, 'the Friend of God' is called Issa Peigamberin (Issa the Prophet, normally a reference to Jesus), and this account evidently encompasses not only the Koranic text of filial sacrifice, but also its biblical counterpart. The name of the figure who is to slaughter iterates, in a peculiar way, the name of the figure who, according to the book of Genesis, is to be slaughtered.

Analysis of the content of Islamic folklore legends of filial sacrifice indicates that these texts relate predominantly, but not solely, to the classical account of the Sacred Book of 'those who have submitted' ('muslimuna'). Indeed, Mohammed's account does not appear to be their ultimate source, but just the major component in the process of shaping the plot of the text. As for the storytellers, needless to say, they regard the Koran as the epitome of their confessional identity. Meanwhile, as they recount it, the story of Abraham has relevance not only for the Koran-based teachings of Mohammed, but to the conventional confessional patterns of 'Folk Islam' as well. Moreover, their narratives refer also to the biblical (either canonical, or apocryphal) account, and to rabbinical or midrashic tradition. At the same time, each storyteller is convinced that his tale is 'the true one', believing that it springs from the Holy Koran itself, and repeats the sacred utterance of Mohammed word for word. And indeed, as E. Leach has emphasised: 'that is how mythologies are presented. They do not exist as single stories but as clusters of stories which are variations around a theme.'[30]

It was also Leach who developed the methodology employed in the present essay, the methodology of the functionalist social anthropologist who believes:

> that sacred texts contain a religious message which is other than that which can be immediately inferred from the manifest sense of the narrative. Religious texts contain a mystery; the mystery is somehow encoded in the text; it is decodable. The code, as in all forms of communication, depends upon the permutation of patterned structures. The method of decoding is to show what persists throughout in a sequence of transformations.[31]

It is my conviction that the folklore narratives concerning the sacred ritual of the 'Kurban' which have been recorded among Muslims in the Balkans also contain an 'encoded mystery'. If deciphered, they could elucidate – both syntagmatically and paradigmatically – the patterned structures that have undergone a series of hidden processes of permutation and survived through the centuries in a sequence of socio-political and cultural transformations.

As for the functional parameters of folklore legends of the Great Trial, they remained constant. However re-narrated, the story of Abraham justifies the main custom of Muslim communities: the annual ritual sacrifice of the lamb or ram and the end of the fast, on the feast day of Ramadan. Their narrative string consists of a series of episodes framing the actions of the protagonists on two hierarchically different levels – terrestrial (represented by Ibraim, and Ismail) and celestial (God, His messenger Dzhebrail/Gabriel), and the ram descending from above to replace one of the earthly protagonists). There is also an antagonist involved in the story about the origin of the ritual of Kurban-Bayram – the Sheitan. Most significant in this respect is the fact that he appeared before Ismail in a marginal place and at a liminal time: the encounter took place on the road, during the journey of the boy from the house to the forest, where the sacrificial offering was to take place. Thus the folklore story implicitly recalls the connotations of the classical ritual pattern of the initiation (from an anthropological point of view), thus referring to the universal cultural paradigm of the rites of passage (and/ or birth/rebirth ceremonies). Moreover, it is a *sui generis* exemplification of how the mytheme of filial sacrifice is realised in the act of performance of the folklore narrative.

Furthermore, as the folklore texts recorded among Muslims in the Balkans indicate, the interaction between the two levels of the

Universe – the terrestrial world and the celestial realms – could be accomplished either by a prayer (an invocation from below to above) or by the ritual of Kurban-Bayram (God's response from above to below). At the same time, these actions of 'terrestrial' subjects involved in the narrative concerned develop predominantly at the horizontal spatial level – apparently with only one exception, when the culmination takes place, and Ismail is replaced by the ram. Otherwise, the plot of the story follows a strict 'horizontal' development. It progresses from the centre of Ibrahim's world ('before the Great Trial') – i.e. from inside his house towards the outside, then to the forest, and then back to the village. Yet neither Ibraim nor his village remains the same, for he brings a new law – the law of the sacred ritual revealed to him during the Great Trial. And finally, the period after the Great Trial marks the final stage of the spiritual metamorphosis of Ibrahim and his village. His world turns out to be a newly reborn and rearranged Universe, through the law of Kurban-Bayram.

On the other hand, the deeds of celestial subjects appear to be accomplished in a vertically framed perspective: with God's heavenly Throne on top; below it, where 'Dzhebrail' (Gabriel) functions as a mediator; and yet further down, the ram, seen as 'betwixt and between'; at the base stands the earth. Thus the place where Ibraim stops to slay Ismail appears to be considered the earthly counterpart of God's heavenly Throne. This is where the ram and Ismail meet – yet now, supposedly, as two equal and therefore interchangeable participants in one cosmological drama. The first descends from above, while the second ascends from below. Thus they represent not only the terrestrial world, but the celestial realms as well, and, what is more important, they spell out the encounter between them. What is more, they also create a precedent, thus revealing the modus operandi of the ritual of Kurban-Bayram as a strategy for making the world of Muslims an earthly counterpart of God's Heavenly Throne. In this way, through the strategy of the ritual, every village can be transformed into Mecca itself.

Along with the explanation regarding the origin of the feast of 'Ram-Bayram' as a sacred precedent established by the prophet Ibrahim, there is one more significant detail in these texts. According to the oral interpretation of the Koranic version of Abraham's sacrifice, Ishmael is considered to be a personification of a sacred genealogy, and an ancestor of the religious community of Islam.

Thus, the Muslim interpretation of the saga of Abraham portrays him as the first one amongst the men 'who had submitted', while the Kurban-Bayram ritual repeats what happened to him on the sacred threshold of 'the new beginning'. That is indeed the nature of myth: 'a sacred tale about past events which is used to justify social action in the present'.[32] It is particularly significant in this respect that, according to one of the storytellers from the Rhodope mountains, the Great Trial is believed to have taken place in his village: 'on that hill over there'.[33] So instead of biblical or Koranic toponyms there appears Bulgarian geographical landscape. Thus the saga of Abraham not only acquires a conventional setting familiar to its listeners, but also transforms the storyteller's neighbourhood into a blessed territory. So the Bulgarian space and therefore Bulgarian toponymy acquire a sacred status. Bulgarian land becomes the Holy Land. The local hill where the Great Trial is believed to have taken place, turns into the Heavenly Altar. At the same time, the folklore text reveals – in terms of ethnopoetics – the symbolic dimensions of the 'altar' concept. The Latin lexeme 'altare' is a derivate from 'altus', meaning 'high', i.e. 'that which is higher than the rest of the temple'.[34] Thus, according to the legend of filial sacrifice, as recounted in the Pirin mountains by the Bulgarian Pomaks, the world of the storyteller encompasses the entire Universe and becomes its sacred centre, in other words its 'Altar'.

It is obvious that, according to the data of folklore versions of the Koran which were recently recorded in the Balkans, the image of Abraham (Ibrahim, Ibraim, Ibriam) personifies obedience to Allah. In a similar way, in the folklore tradition of Christians, the image of Avram also epitomises obedience to God. Thus both Muslims and Christians use the same character – Abraham – in order to spell out the paradigm of the 'true faith'.

Conclusions

The folklore interpretations of the saga of Abraham, as recorded in Muslim communities in the Balkans, offer convincing evidence to corroborate the theoretical presumptions of Haim Schwarzbaum: 'There is no doubt that a folkloric approach to the Koran's extremely interesting stories does more to advance our knowledge of the Holy Book of Islam than many a theoretical treatise ignoring the intricate process of oral narration.'[35]

The interdependence between folklore and Christian canons, on the one hand, and oral tradition and Muslim traditions on the other, continues to be an active process in Balkan oral tradition. Reflecting folk perceptions of Christian and Muslim material, the popular interpretation of the saga of Abraham illuminates the empirical image of the Holy

Scriptures in their folklore dimensions. It also illustrates some aspects of the confessional core of Balkan traditional culture. It reflects a centuries-old oral tradition, complementing or acting as an alternative to Old and New Testament canonic (and apocryphal) texts, and to the Koran. Further, it captures different stages in the tradition's development.

Although they were recorded recently, it is possible to argue that some of the folklore materials I have mentioned predate the transition of biblical or Koranic proto-texts into written text. As such, they hold a particular fascination. In fact, folklore was not only a universal source of ritual and narrative religious paradigms, but also a cradle for the emergence both of the Bible and of the Koran.

Notes

1 This chapter incorporates some results of my research project analysing parameters of folk religion in Slavonic and Balkan traditional culture on which I am currently working at the School of Slavonic and East European Studies, University of London. The project, entitled *The Folk Bible: Bulgarian Oral Tradition and Holy Writ*, concentrates on the description of popular dimensions of Christianity and Holy Writ in particular. The work is based on texts I recorded in field research over the last two decades. The study was accomplished with the assistance of the Modern Humanities Research Association.

2 See in this respect O. Andersen, 'Oral Tradition', in H. Wansbrough (ed.), *Jesus and the Oral Gospel Tradition. (Journal for the Study of the New Testament*; Supplement Series 64), Sheffield, JSOT Press, 17–58; D. Aune, 'Prolegomena to the Study of Oral Tradition in the Hellenistic World', *ibid.*, 59–106; D. Damrosch, *The Narrative Covenant. Transformations of Genre in the Growth of Biblical Literature* (San Francisco: Harper & Row, 1987), 10, 155, 233–41; W. Kelber, *The Oral and the Written Gospel. The Hermeneutics of Speaking and Writing in the Synoptic Tradition, Mark, Paul and Q.* (Philadelphia: Fortress Press, 1983); P. Kirkpatrick, 'The Old Testament and Folklore Study', *Journal for the Study of the Old Testament*. Supplement Series 62, Sheffield: JSOT Press, 1988; B. Lang (ed.), *Anthropological Approaches to the Old Testament. (Issues in Religion and Theology Series* 8), (Philadelphia: Fortress Press & London SPCK, 1985); S. Niditch, *Folklore and the Hebrew Bible* (Minneapolis: Fortress Press, 1993); J. Rogerson, *Anthropology and the Old Testament* (Oxford: Basil Blackwell, 1978); H. Ruger, 'Oral Tradition in the Old Testament' in Wansbrough, *op. cit.*, pp. 107–20; H. Schwarzbaum, *Biblical and Extra-Biblical Legends in Islamic Folk-Literature* (Verlag für Orient-kunde Dr. H. Vorndran, Walldorf-Hessen, 1982), pp. 10–60; M. Soards, 'Oral Tradition Before, In, and Outside the Canonical Passion Narratives', in Wansborough, *op. cit.*, pp. 334–50; E. J. Van Wolde, 'A Semiotic Analysis of Genesis 2–3. A Semiotic Theory and Method of Analysis Applied to the Story of the Garden of Eden', *Studia Semitica Neerlandica* (Maastricht: Van Gorcum, 1989).

3 On the theological interpretation of the saga of Abraham, see A. S. Herbert, *Genesis 12–50* (London: SCM Press Ltd., 1962); and J. Janzen, *Abraham and All*

the Families of the Earth. A Commentary on the Book of Genesis 12–50 (Edinburgh: Handsel Press Ltd., 1993); on the types of discourse which characterise the Genesis narratives concerned with Abraham and his offspring, see White (1991): 148–273; on the textual interpretation of the Biblical text about the ordeal of Isaac see D. Greenwood, *Structuralism and the Biblical Text* (Religion and Reason Series 32: Method and Theory in the Study and Interpretation of Religion) (Berlin, New York, Amsterdam: Mouton Publishers, 1985), pp. 123–9.

4   A. Niccacci, 'On Hermeneutics of the Holy Scriptures', in F. Manns (ed.), *Studium Biblicum Franciscanum Analecta 41. The Sacrifice of Isaac in the Three Monotheistic Religions. [Proceedings of a Symposium on the Interpretation of the Scriptures held in Jerusalem, March 16–17, 1995]* (Jerusalem: Franciscan Printing Press), pp. 177–80.

5   The Biblical formula 'as the stars of the heaven' occurs in particular folklore genres (panegyrical chants, spells and blessings), usually marking the climax of their performance.

6   This text was published in 1891, in the *SBNU* (*Sbornik za Narodni Umotvorenia*) Bulgarian Folklore Collection series, vol. 6, p. 80 (text no. 2). It is the formulaic climax of a song sung in Eastern Bulgaria (the area of Shumen and Tărnovo). See also N. Kaufman, *Narodni pesni na Bălgarite ot USSR & MSSR. Tom 1–2* (Sofia: Izdatelstvo na Bălgarskata Akademia na Naukite, 1982), p. 469, text 699.

7   I. Duychev, *Păteki ot Utroto. Ochertsi za Srednovekovnata Bălgarska Kultura* (Sofia: Otechestvo, 1985), p. 97.

8   The text entitled 'Sermon on Righteous Abraham' (Slovo o Pravednago Avraama) was copied in 1628 by the Bulgarian monk Daniil from the Holy Trinity Monastery in the vicinities of Etropole (in the Balkan ranges); for further information see F. Badalanova and A. Miltenova, 'Apokrifniat Tsikăl za Avraam văv Folklora i Srednovekovnite Balkanski Literaturi', in *Etnografski Problemi na Narodnata Kultura*, vol. 4 (Sofia: Etnografski Institut s Muzei pri Bălgarskata Akademia na Naukite, 1996), pp. 203–51.

9   J. Hanauer, *The Holy Land – Myths & Legends* (first published in 1907 as *Folklore of the Holy Land* by The Sheldon Place, London) (London: Senate), p. 23.

10 See the texts of some ritual songs of Abraham's sacrifice published in the first volume of the two-volume *Bălgarski Narodni Baladi i Pesni s Miticheski i Legendarni Motivi* anthology of Bulgarian folk ballads produced by the Institute of Folklore at the Bulgarian Academy of Sciences, Sofia as a separate issue (no. LX) of *SBNU* Collection: L. Bogdanova, S. Boyadzhieva, N. Kaufman, K. Mikhaylova, L. Parpulova, S. Petkova and S. Stoykova (eds), *Bălgarski Narodni Baladi i Pesni s Miticheski i Legendarni Motivi. Sbornik za Narodni Umotvorenia i Narodopis*, kn. LX. 1–2 (Sofia: Izdatelstvo na Bălgarskata Akademia na Naukite, 1993–4), pp. 364–373, and especially text no. 485 (recorded by L. Bogdanova near Yambol, South-Eastern Bulgaria); text no. 486 (recorded by L. Bogdanova near Silistra, North-Eastern Bulgaria); text no. 489 (recorded by L. Parpulova in the same area).

11 N. Nachov, 'Tikveshki Răkopis', in *Sbornik za Narodni Umotvorenia, Nauka i Knizhnina. Izdava Ministerstvoto na Narodnoto Prosveshtenie*, kn. X (Sofia: Dărzhavna Pechatnitsa, 1894), p.150.

12 All these forms relate, of course, to the Old Church Slavonic агna, агьць, trNьць агNьць, etc. These lexemes, in turn, are considered derivates from the Indo-European *ag(h)no-s (see also Lat. agnus).

13 See Dimitri and Konstantin Miladinovtsi, *Bălgarski Narodni Pesni Sobrani ot Bratia Miladinovtsi Dimitria i Konstantina i Izdani ot Konstantina v Zagreb v Knigopechatnitsata na A. Iakitcha* (Zagreb: A. Iakich, 1861), text no. 29.

14 See P. Bessonov, *Kaliki Perekhozhie. 1. Vyp. 3* (Moskva, 1863), text no. 135, entitled *Боlгарский Духоvнй cтих "Авраамоvиць"* ['The Avraamovitsy' Spiritual Stanza].

15 See V. Kachanovskii, *Sbornik Zapadno-Bolgarskikh Pesen* (St Petersburg, 1883), text no. 41 and text no. 42.

16 See N. Bonchev, *Sbornik ot Bălgarski Narodni Pesni* (Varna, 1884), text no. 12.

17 There are numerous texts published in various collections of South-Slavonic folk songs, such as: B. & M. Angelov, Arnaudov, *Bălgarska Narodna Poezia*, Vol. 1 (Sofia, 1923), p.146, text no. 19; B. Angelov & Ch. Vakarelski, *Senki iz Nevidelitsa* (Sofia, 1936), pp. 221–4, text no. 74; M. Arnaudov, *Bălgarski Narodni Pesni*, vol. 3: 'Nuveli, Baladi i legendi' (Sofia, 1939), text no. 44; L. Dumba & J. Jovanović, *Epske Pjesme is Makedonije*, IV, 3, 1913, text no. 3; L. Karavelov, *Bolgarskyia Narodnyia Pesni. Izdannyia pod Nabliudeniem P. A. Lavrova* (Moskva, 1905), text no. 134; Kaufman, *op. cit.*, text nos 1459, 1460; V. Krasić, V., *Srpske Narodne Pjesme* (Pančevo, 1880), text no. 15; P. Mikhailov, *Bălgarski Narodni Pesni ot Makedonia* (Sofia, 1924), text no. 239; M. Obradović, *Srpske Narodne Ženske, Djevojačke, Lirske Pjesme iz Slavonije* 1890–1, text. no. 118; B. Rusić, *Prilepski Guslar Apostol* (Beograd, 1940), text no. 1; V. Stoin, *Narodni Pesni ot Sredna Severna Bălgaria* (Sofia, 1931), text no. 141; and *Bălgarski Narodni Pesni ot Iztochna i Zapadna Trakiia* (Sofia, 1939), text nos 78, 79; see also texts published in the *SBNU [Sbornik za Narodni Umotvorenia]*, Bulgarian Folklore Collection series: vol. 1: p. 27, text no. 4; *SBNU*, vol. 2, p. 22, text no 1; p. 23, text no. 2; p. 24, text no. 3; p. 25, text 4; *SBNU*, vol. 3, p. 38, text 1; *SBNU*, vol. 10, p. 11, text 3; *SBNU*, vol. 27, p. 302, text 211; *SBNU*, vol. 35, p. 336, text 406; *SBNU*, vol. 36, p. 147, text 146, p. 162, text 150, p. 163, text 151, p. 163, text 152; *SBNU*, vol. 43, pp. 252–62; as well as A. Stoilov's Motif-Index of Bulgarian folk songs recorded during the nineteenth century: A. Stoilov, *Pokazalets na Pechatanite prez XIX Vek. Bălgarski Narodni Pesni*, vols 1–2 (Sofia, 1916–18) vol. 1 [1916: 49] & vol. 2 [1918: 568]; B. Krstić's Motif-Index of the folk songs of the Balkan Slavs: B. Krstić, *Index Motiva Narodnih Pesama Balkanskih Slovena* (Beograd, 1984), 603; and the two-volume *Bălgarski Narodni Baladi i Pesni s Miticheski i Legendarni Motivi* anthology of Bulgarian folk ballads published by the Institute of Folklore at the Bulgarian Academy of Sciences, Sofia as a separate issue (no. LX) of *SBNU* Collection: Bogdanova et al. (eds), 1993, 548–51. On the folklore dimensions of the 'Kurban' ritual, as performed amongst the Balkan Slavs, see A. Popova, 'Le Kourban, ou Sacrifice Sanglant Dans les Traditions Balkaniques', in *Europaea. Journal of the Europeanists*, 1995, I–1, pp. 145–70 (Cagliari & Bruxelles: Instituto Superiore Regionale Etnografico & The Europeanists Society).

18 Indicative in this respect is a version registered in 1880 in the village of Kokre by one of the local intellectuals and ecclesiastical leaders A. Ianov who, on the day when the 'Kurban' ritual was to be performed, met with the story-

tellers of the neighbourhood. According to his report, the purpose of his visit was to convince them that the 'Kurban' ritual was not a Christian, but a heathen feast. Apparently, they were unwilling to participate in the dispute and refused to listen to him; instead, they started singing the ritual song of Abraham's sacrifice, and, as Ianov summed it up, 'it was their ultimate answer to the matter in question'. Thus, according to the local storytellers Dimitrii Peykov, Dimitrii Trupkov and Iove Naydenov, the song of Abraham's sacrifice must be sung by men on the Pentecost feast day, during the 'Kurban' ritual. Ianov's account was eventually published by the Russian scholar Mikhail Selishchev in his monograph on the culture, language and folklore of the Christian settlement in the Polog neighbourhood (the Moriovo area). For further details, see A. Selishchev, *Polog i ego Bolgarskoe Naselenie. Istoricheskie, Etnograficheskie i Dialektologicheskie Ocherki Severo-Zapadnoi Makedonii [S Etnograficheskoiu Kartoiu Pologa]* (Sofia, 1929), pp. 207–8.

19 The song was peformed by a 66-year-old Christian woman by the name of Stana Bozhkova Vlaeva, born and married in the same village, unschooled, a farmer. The text is to be sung by women during the 'Kurban' ritual, on the Gergiovden Feast Day (the Feast Day of St George), when the priest blesses the loaves of ritual bread and scents the roasted lamb. See also text no. 491 and text no. 493, published in the first volume of the *Bălgarski Narodni Baladi i Pesni s Miticheski i Legendarni Motivi* anthology of Bulgarian folk ballads, Bogdanova et al., 1993, pp. 370–2.

20 A similar approach to the 'wood'/'forest' topic is to be found in some Bulgarian ritual songs; see, for example, the account given to K. Mikhaylova in the village of Popina in the vicinities of Silistra, by a 61-year-old Christian woman by the name of Yordana Dragostinova, as well as the account from the village of Gălăbintsi, Yambol area, given to L. Bogdanova by Dina Mikhaylova, a 58-year-old woman. The texts in question are published in the first volume of Bogdanova, *op. cit.*, pp. 364–5.

21 This analysis is based upon a corpus of texts recorded by my colleagues from the Institute of Folklore (Bulgarian Academy of Sciences), Sofia. I am deeply indebted to Evgenia Mitseva, Milena Benovska and Tanya Boyadzhieva who allowed me to refer to the results of their field-research amongst the Muslim communities in Bulgaria. For further information see F. Badalanova and A. Miltenova, *op. cit.*, pp. 246–9.

22 On the interpretation of the theme of the sacrifice of Isaac in Early Patristic exegesis see L. Cignelli, L. 1995, 'The Sacrifice of Isaac in Patristic Exegesis', in *Studium Biblicum Franciscanum Analecta*, 41 (1995) (F. Manns (ed.), *The Sacrifice of Isaac in the Three Monotheistic Religions. Proceedings of a Symposium on the Interpretation of the Scriptures*, held in Jerusalem, 16–17 March 1995) (Jerusalem: Franciscan Printing Press, 1995), pp. 123–6, and Paczkowski, *ibid.*, pp. 101–21.

23 Doukhan 1995, p.167.

24 See text no. 491 (recorded by L. Parpulova in 1980 in the vicinities of Kazanlăk), text no. 492 (recorded by S. Boyadzhieva in 1980 in the vicinities of Stara Zagora), text no. 493 (recorded by L. Bogdanova in 1976 in the vicinities of Kharmanli) and text no. 494 (recorded by L. Parpulova in 1980 in the vicinities of Russe), published in the first volume of Bogdanova et al., *op. cit.*, pp. 370–3.

25   S. Kunin, 'The Death of Isaac: Structuralist Analysis of Gen. 22.', in Manns, *op. cit.*, p. 54.
26   *Ibid.*, p. 55.
27   *Ibid.*
28   J. Gutmann, *Hebrew Manuscript Painting* (London: Chatto & Windus, 1979), p. 11.
29   This type of ritual song also correlates with some legends, widespread amongst the Southern Slavs. According to them, the Lord comes down to Earth as a stranger (a beggar, an old man, etc.) staying in the house of a righteous Christian, whose only wealth is his one and only son. The poor peasant who did not wish to disgrace himself before his guest, received Him with respect. As a 'God-fearing man', he offers Him a 'worthy dish'; and since he has no sheep, instead of a roasted lamb he places on the low table another meal – his child, roasted on a plate. The Lord, however, works a miracle and the offspring of the pious Christian is resurrected. For further information see M. Dragomanov, 'Slavianskite Skazania za Pozhertvuvanie Sobstvenno Dete', in *Sbornik za Narodni Umotvorenia, Nauka i Knizhnina*, Izdava Ministerstvoto na Narodnoto Prosveshtenie, kn. 1 (Sofia: Dărzhavna Pechatnitsa, 1889), pp. 65–96.
30   E. Leach, and D. A. Aycock, *Structuralist Interpretations of Biblical Myth* (Cambridge: Cambridge University Press, 1983), p. 2.
31   *Ibid.*
32   *Ibid.*, p. 8.
33   The text was recorded in the village of Ribnovo, Gotse Delchev area, by the Bulgarian folklorist Stoyanka Boyadzhieva. The legend was narrated by Ilia Likov, a 57-year-old mountain dweller, a Pomak.
34   M. Fasmer, *Etimologicheskyi Slovar' Russkogo Iazyka. Perevod s Nemetskogo i Dopolnenia Chlena-korrespondenta Akademii Nauk SSSR O. N. Trubachova. Pod Redaktsiei s Predisloviem Prof. A. Larina. Izdanie vtoroe, stereotipnoe. V chetyrekh tomakh*, vol. I (A–E) (Moskva: Progress, 1986), vol. 1, p. 72.
35   Schwarzbaum, *op. cit.*, p. 12.

3
Art and Doctrine in the First Bulgarian Kingdom

Elissaveta Moussakova

Christianity became established in the First Bulgarian Kingdom between 864 (866) AD and 1018, when the kingdom fell under Byzantine rule.[1] The purpose of this chapter is to present several typical examples of the link between art and doctrine that highlight the specific nature of the 'Bulgarian transition' from paganism to Christianity.[2] These occurred in the period of existence of the autocephalous Bulgarian Church and in the proximity of the two ancient Bulgarian capitals, Pliska and Preslav, in which fairly compact groups of monuments from the conversion period have survived.

First, a few facts of history to set the scene: in the lands of the First Bulgarian Kingdom, the conversion to Christianity probably stemmed from a religious pluralism based on Bulgar and Slav traditions which have been described as tending to monotheistic doctrines[3] and evidently not hostile to each other. Another factor common to both ethnic elements may have been the practice of 'shamanism'.[4] There is no conclusive evidence to suggest the existence of a single state religion or religious authority vested in a particular institution or the dominance of either the Bulgar or the Slav pantheon with their respective supreme gods Tangra and Perun.[5] On the other hand, the remains of pagan places of worship, subsequently converted into Christian temples, suggest a certain organisation of religious life and practice, while some historical sources describe the Bulgarian ruler as a High Priest officiating at particular rituals.[6] It is also quite likely, judging by the forms taken by early Bulgarian Christianity, that the element of cohesion between the two ethnic faiths in the kingdom was provided by the cult of the ruler, the founder of the state and, hence, of ancestors whose expression in rituals remains virtually unknown to us.[7] Some pagan rulers, such as Omourtag, who went down in legend as a fierce persecutor of Christians in his

lands, adopted many of the outward forms of Byzantine imperial doctrine, the theological and political foundations of which had been laid
by Emperor Constantine the Great.[8]

The period in which Prince Boris I of Bulgaria baptised his people
saw the defeat of iconoclasm, and the act of conversion itself was
accompanied by political manoeuvring between Rome and Constantinople.[9] The Ecumenical Council of 870 in Constantinople resolved the
controversy by placing the Church of Bulgaria under the jurisdiction of
the Church of Constantinople, granting it the authority of an autocephalous archbishopric. Boris's successor King Simeon upgraded that
status to a patriarchate (927) in defiance of objections raised by the
Byzantine Patriarch. It was not until some time during the reign of
Simeon's son Peter that the then Emperor of Byzantium, Romanus
Lecapenus, granted official recognition to the Bulgarian Patriarch,
while, for all practical purposes, Byzantium continued to regard the
independence of the Church of Bulgaria as illegitimate.[10] Preslav was
the seat of the Bulgarian Patriarch until it was captured in 976 by the
troops of Emperor John Tzimisces, when the centre of church authority
moved to Dorostol (now Silistra) on the Danube but as an archbishopric
subordinate to Constantinople.[11]

With the establishment of Christianity as an official religion in the
Roman Empire, art was called upon to create the visual image of the new
doctrine within the necessary dogmatic and theological frame of reference that would distinguish the Christian icon from the pagan idol.
Once the heresy of iconoclasm had been defeated, the use of images
became firmly established, with their potential to express visually the
fundamental themes of Christianity. That was the basis for the development of a body of images that were generally accepted, generally recog-
nisable and approved by the Church, and carried a definite ideological
message. The latter co-existed in a more or less parallel fashion with a
legacy of traditional beliefs, inherently pagan, which the Church and its
missionaries sought to incorporate into, or reconcile with, the new
Christian dogma. That legacy, too, found its own visual form, which
could be described as marginal to the official iconography and represen-
tational rather than conceptually expressive. On the other hand, it was
an element in a far broader paradigm that brought together Christian
dogma, pagan belief, ethnic tradition, apocrypha and heresy. The few
scattered cultural monuments that remain of the two Bulgarian capitals
of the ninth and tenth centuries do allow an analysis, however brief
and fragmentary, of some correlations between Christianity and paganism, official doctrine and 'vernacular' Christianity, an iconographic

programme and the spontaneous symbolism that constitute the pattern of the 'Bulgarian case' of Christianisation.

Doctrine and official art

What remains of Pliska, the capital which saw the conversion of Prince Boris and his court, are architectural monuments that revived or copied features of early Christian basilicas. This form provided sufficient space to accommodate the large numbers of newly baptised in a massive process of conversion, while it also corresponded to the Church's main doctrine. Some scholars have attributed some of the earliest Bulgarian basilicas to the influence of the Latin Mission of 866–70, which must have brought with it models of Western sacred architecture.[12]

Christian art of this phase, the spread of Christian precepts, ethics and dogma, flourished during the reigns of King Simeon and King Peter in Preslav, which had become the capital after the so-called 'heathen rebellion' lead by Vladimir, Simeon's brother and heir apparent to the throne of Boris, who was dethroned and blinded when the rebellion was crushed. The ruins of the building and fragments of decoration, such as stone sculpture and painted ceramics, and the overall cultural context suggest beyond any doubt that the Bulgarian ruler aspired to build a capital that would rival Constantinople.[13] There seems to be more than mere romantic nostalgia in the idea that King Simeon played a key role in organising the institutions of the newly adopted faith and encouraged the artistic endeavour that ultimately created a symbolic order inspired by a set of ideological values. Admittedly, the context of this discourse on early Bulgarian art is far from complete, but what has survived and the parallel with post-iconoclastic Byzantine art suggest an idiom that was more ornamental than figurative. It abounded in zoomorphic imagery and floral motifs, while the surviving traces of anthropomorphic representation, notably painted ceramic icons, do not seem to have formed part of any complete iconographic cycle but were rather isolated 'portraitures'. Studies of iconoclastic and post-iconoclastic Byzantine art have concluded that new developments in imperial doctrine led to an unprecedented proliferation of the ornament, the ideogram, moving away from what was regarded as idolatrous mimesis towards a decorative luxury of oriental proportions.[14] When iconolatry was finally re-established, this doctrine evolved into a 'neo-Constantinian' apotheosis of the ruler as a warrior of Orthodoxy. In my view, it was this particular context that nurtured, in a rather curious way, newly converted Bulgarian culture.

Stone pillars

If archaeological studies can be relied upon, one category of monuments dating from pagan times continued to function at the level of official culture after the conversion to Christianity. These were stone pillars bearing inscriptions of various content but mostly related to the military affairs of pagan Bulgar rulers.[15] Some of them, such as those dating from the reign of Khan Kroum, were discovered in the interior of the Great Basilica, the most important monument in Pliska.[16] The famous pillar in the monastery at Ballsh (now in Albania), bearing an inscription about the conversion, survives from the time of Boris I,[17] and the pillar from Narush (now Nea Philadelphia in Greece) served as a border post during the reign of Simeon.[18] The practice of erecting such pillars was so deeply rooted that it was recorded centuries later, when, in 1230, King Ivan-Asen II placed Omourtag's memorial pillar (by that time, already a piece of ancient heritage) in the interior of the Church of the Holy Martyrs of Sebastia in Turnovo built and decorated to celebrate the King's victory over Emperor Theodore Comnenus.[19] Apparently, the pagan form of celebrating and commemorating the ruler's triumph in battle had survived as a symbol of sovereign continuity, because it had followed a Byzantine, Christianised tradition, much as Byzantium had inherited its triumphal forms from Rome.

The Madara Horseman

The famous rock relief, whose age and meaning are still the subject of much debate, was undoubtedly a cult image in the ancient, possibly Thracian, complex of sacred structures at the foot of the Madara plateau (not far from the capital Pliska).[20] Archaeological evidence supports the view that, shortly after the conversion, churches were built close to and on top of pagan shrines, but the relief carved in the face of the rock overlooking this site was preserved.[21] If the Horseman represented the Bulgar supreme god Tangra, as some Bulgarian scholars have argued, then its survival dominating the new Christian temples would be most surprising.[22] I am more inclined to believe that the monument was preserved because its meaning lay in the cult of ancestors, the cult of the progenitor, and the Christian ruler could easily be associated with it as he himself repeated the deeds of his predecessors, albeit in a new key. He was a warrior and a Baptist and, thus, a giver of new life to his subjects.[23] The worship of the Madara Horseman as St George does suggest, in my view, the kind of discourse that was applied to the semantic transformation of pagan doctrine.[24] As a colleague, Oksana

Minaeva, has observed, the worship of the pagan progenitor as none other than a horseman was entirely consistent with the socio-cultural environment of the Bulgars and, I would add, the Slavs who co-existed with them.[25] There is a literary source, known as The Tale of the Iron Cross, which survives in a latter-day version,[26] but does exhibit some traces of an ancient Bulgarian substratum.[27] In it, the hero's adventures and his delivery through the intercession of St George take place in a setting that is indicative of a warrior cult clearly based on a pagan initiation archetype.[28]

This hypothesis that the pagan relief of the horseman came to be venerated as a popular Christian saint thanks to the formal resemblance between the two is unfortunately shattered by a different interpretation of the archaeological evidence. According to Stojan Maslev the ruins of the churches at the foot of the Madara rocks date from the thirteenth–fourteenth centuries and not from the time of conversion.[29] The same holds true for the cave colonies in the region carved, it seems, by hesychasts. As S. Maslev was a reliable observer, we must rather think about a time of neglect of 'the greatest shrine of the First Bulgarian Kingdom'.[30] It might have been centuries later, when paganism was no longer a threat, that the place was inhabited by a monastic community. It is possible, then, that the veneration of the Madara horseman as St George did not begin before the Ottoman conquest of Bulgarian lands, in an epoch of searching for evidence of ancient, already forgotten, traditions, which gave the people the strength to survive.

These examples, however disputable, suggest one way in which the early Christian doctrine of the ruler was permeated by a pagan undercurrent, which was of vital importance for the preservation of Bulgaria's cultural identity, given the constant threat of political and cultural assimilation by Byzantium.

The cross: the symbol of paganism defeated

As in early Christian times, during the iconoclastic and post-iconoclastic periods, the doctrine of the cross was raised to a new height of importance. It symbolised the fundamental Christian doctrine of Salvation and both the temporal and the spiritual authorities resorted to it at times of crisis or change in the life of the Church. The cross-domed church of the post-iconoclastic period, which came to Preslav in the tenth century, was undoubtedly created as one of the instruments symbolising the victory of Orthodoxy. For a newly converted country such as Bulgaria, it also manifested the establishment of the new faith through its form of elevated cross.

Since Constantine the Great, the cross had become established, as André Grabar has put it, as an instrument of the Emperor's triumph and a symbol of secular Christian authority.[31] While the answers that Pope Nicholas I gave to the Bulgarians' enquiries leave us in no doubt that the doctrine of the cross was expounded to the new proselytes with catechetical precision, there was a curious nuance that is symptomatic of the technique of conversion: while the Pope emphasised the spiritual meaning of the cross, he did not fail to make the symbol more understandable by drawing an analogy with something familiar – the ancient Bulgar battle standard, the horsetail.[32] Thus, the cross took on the significance of a weapon of war, which it bore ever since as a hidden hallmark. This theme probably featured in the early sermons, side by side with traditional, but probably rather emphatic, interpretations of the cross as 'a weapon against the temptations of the Devil', which are ubiquitous in the surviving texts of early Bulgarian theologians. It is thus quite likely that, in the Bulgarian pagan mind, and for the purposes of its conversion, the Christian doctrine was clad in warrior/ruler rhetoric. Incidentally, this strategy was typologically common to all barbarian peoples in Europe whose main trade was war.[33]

In the realm of official art, there was apparently no visual expression of the symbol's ambivalence. On the contrary, the evidence suggests that it was rendered in a rather traditional fashion and use was quite often made of the surviving sculptured fragments of early Christian temples, which contained typical compositions of crosses and floral motifs. A distinctive feature of the iconography of the cross in Byzantium and in Bulgaria was the motif of the leaved cross,[34] the emblem of triumphant iconolatry[35] and, in the case of Bulgaria, of triumphant Christianity itself. The various renditions of the cross in old Bulgarian art were a matter of semantic overtone within the framework of the Christian doctrine. Things were different, however, in the realm of marginal culture.

The doctrine in a lesser guise

The artistic specimens that readily come to mind are the drawings, or graffiti, scratched on church and castle walls – a form of artistic expression that has accompanied every 'high' culture.

The magic power of the cross

I do not propose to dwell at any length on the transposition of the Christian cross to the level of folklore as a powerful apotropaic object;

this is a well-known fact. Among the graffiti, found in particularly large numbers on the castle walls of Pliska and Preslav – the southern portions of the Round Church in Preslav and, especially, the exterior walls of the monastery at Ravna – the cross is by far the most frequent motif and this is quite natural. Moreover, in contrast to the conventional forms of the cross in official art, the graffiti exhibit a considerable variety of both form and composition, including other symbols and images: geometrical, floral, zoo- and anthropomorphic. Quite often the cross is inscribed in a circumference of dots to denote its special power. It is of course impossible, in a number of cases, to distinguish archaeologically between drawings that date from pagan times and those that date from Christian times. However, the abundance of compositions involving the cross does suggest the activating role of the Christian sermon, under the influence of which the symbolic archetypes of the pagan tradition acquired new meanings. It was, I believe, also not until that influence came into play, that the visual equivalents of that tradition became more distinct. Among the most interesting examples are those from the monastery at Ravna.[36]

The most common form is the 'cross fourchée' (or fork-shaped) with arms divided at the ends. It seems that this was the common graphic representation of the 'cross patoncée' described by A. Grabar.[37] On the other hand, this form correlates with another specific character: upsilon and upsilon in stems (or vertical brackets), which is very common in the Pliska area.[38] Scholarly opinion used to relate that to the Bulgar supreme god Tangra[39] but materials on the basis of which the drawings have been dated suggest the ninth and tenth centuries as the period of peak proliferation and, on the strength of this argument, a different interpretation has been gaining ground.[40] Still, whether the form was the emblem of a Christian sect or, indeed, a pagan apotropaic character denoting the presence of a superhuman force is yet to be decided. It is only worth mentioning here that it has been discovered marking the neck of a ceramic vessel, numerous building fragments, and roof tiles from Simeon's Church in Preslav. Another find, the amulet medallion from Preslav, bears the upsilon in stems on one side and the Christian cross with fork-shaped horizontal arms on the other.[41]

Among the various combinations of the cross fourchée with other images, I shall only dwell in brief on the one with the boot. A young researcher[42] recently proposed an interesting and rather convincing identification of the boot in a Christian, rather than a pagan shamanic context, as it had usually been interpreted.[43] This new hypothesis relates the boot to some scenes in Byzantine illuminated manuscripts, showing the boots of Moses, off his feet, as he had turned to face the burning bush on Mount Horeb, and also, to the biblical symbolism of the shoe denoting the road to God.[44] Hence, the boot is interpreted as a non-canonical symbol of Christ, introduced, probably, by some monastic brotherhood, and the particular form of the boot is attributed to historical realia. We should not ignore the fact that the shoe or, rather, the foot has been widely identified as one of the early Christian symbols denoting pilgrimage or the pilgrim's progress on the road to Salvation. Moreover, in almost all of the cases that I am referring to, the image of a boot is combined with that of a cross, and that, most frequently, is the fork-shaped one, lending credence to the hypothesis of Christ being represented by the road to Him, the road of salvation and triumph over Satan. Could this be the interpretation of the unusual drawing (Fig. 3.1) at Ravna,[45] featuring a cross fourchée, a boot in glory and a

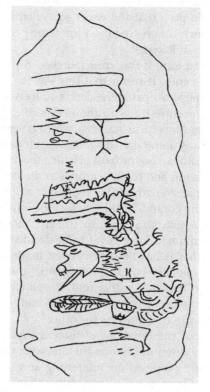

Figure 3.1   Graffiti from Ravna representing two boots, a lion and a cross, ?tenth century (after R. Kostova). Courtesy of Cyril and Methodius National Library.

lion rampant trampled under it? What could be the reference here to the Virgin Mary of the Burning Bush, if the above interpretation is correct and bearing in mind the lettering MARINA (MARINA) inscribed on the boot, as well as another composition, cited by the same author, of a cross and a boot on a tenth-century amphora from Capidava (Romania) inscribed with Virgin Mary's initials?[46] These are all questions which have still to be answered.

The warrior carrying a cross

The final example is from another broad thematic cycle identified in the graffiti: the horseman, either on his own or in hunt or battle scenes. For the Bulgarians, as people who virtually lived on horseback, these images were a vehicle of self-identification. In addition, however, many of them feature the cross and, among the graffiti at Ravna, there are also unarmed horsemen carrying a cross. Their posture is akin to that of an orant and their clothes are hatched, as are those of shamans and saints, which suggests that these were the images of cavalier saints whose miraculous deeds were perceived at the level of folklore by analogy with shamanic practice. The absence of a halo further emphasises the ambivalence of the image. Another drawing at Ravna features a horseman, not carrying either a weapon or a cross, but hanging from the horse's harness is the familiar upsilon character.

In contrast to the Madara Horseman, bearing all the warrior/hunter attributes – an entirely pagan image – the graffiti horsemen carrying a cross exemplify the transformation of traditional concepts from the opposite perspective. Where the two lines of Christian indoctrination intersect is the figure of the horseman which seems to have acted as a mediator, both semiotic and metaphorical, between two different outlooks the specific convergence of which mirrored the Christianisation of the First Bulgarian Kingdom.

Notes

1 P. Aubin, *Le problème de conversion* (Paris, 1963).
2 This paper is based on my PhD thesis 'Promeni v semantikata na njakoi motitvi v izkustvoto na Parvata balgarska darzhava (Prehod ot eziestvo kam christianstvo)', Balgarska Akademija na naukite, Institut za izkustvoznanie (Sofia, 1988).
3 V. Beshevliev, *Parvobalgarite. Bit i kultura* (Sofia, 1981), p. 74 refers to a proto-Bulgarian inscription on stone where the text says that the 'God (of the Bulgars, E.M.) sees'. It seems that at the time of the conversion Perun was venerated as the highest deity in the hierarchically undeveloped religious system of the Bulgarian Slavs. Some authors suppose that an increasing

process of monotheisation was apparent due to contacts with Byzantine culture on the one hand and proto-Bulgarian culture on the other, see A. Stoinev, *Balgarskite slavjani, mitologija i religija* (Sofia, 1988), pp. 23, 29.

4   Here I use the term 'shamanism' in a broader meaning than its classical definition, i.e. M. Eliade, *Shamanismat i archaichnite tehniki na extasa* (Sofia, 1996).

5   As the late Prof. V. Velkov once noted, the situation in the First Bulgarian Kingdom might have been very similar to that in the Roman Empire where paying due honour to the Emperor left people free to venerate their own deities.

6   According to V. Beshevlev, *Parvobalgarski nadpisi* (Sofia, 1979), no. 6, p. 123, the inscription on a small battered column found in a church in Madara states that Khan Omourtag has offered a sacrifice to the god Tangra which does not necessarily mean that the Khan performed the role of priest. The Byzantine historian Theophanes describes the magic rituals performed by Khan Kroum when his troops laid siege to Constantinople, Theoph. 503, 4–14 (*Izvori za balgarskata istorija*, VI., p. 289); also *Scriptor incertus*, 342, 1–15.

7   In her study, O. Minaeva, *Madarskijat konnik* (Sofia, 1990), pp. 110–22, summing up the researches of other scholars, suggests that the notion of the Bulgarian ruler reflected the features of the cultural hero of the Turks who was seen as a horseman.

8   See A. Grabar, *L'Empereur dans l'art byzantin* (Paris, 1936); I. E. Karajanopoulos, *Politicheskata doktrina na vizantijzite* (Sofia, 1992) (Series editionum traductarum. Vol. I).

9   The date of conversion is still disputable in Bulgarian historical studies. See T. Vasilevski, 'Datata na pokrastvaneto na balgarite', in his *Balgaria i Vizantija, IX–XV vek* (Sofia, 1997), 31–42. He himself claims the year 866 while most Bulgarian scholars still accept 864 or 865.

10   I. Snegarov, *Istorija na Ohridskata arhiepiskopija-patriarshija*, I. (Sofia, 1995), pp. 8–9.

11   *Ibid.*

12   N. Chaneva-Dechevska, *Carkovnata arhitectura na Parvata balgarska darzhava* (Sofia, 1984), pp. 24–32.

13   I. Bozhilov, *Zar Simeon Veliki (893–927): Zlatnijat vek na Srednovekovna Balgaria* (Sofia, 1983), pp. 56–7.

14   A. Grabar, 'Religioznoto izkustvo i vizantijskata imperija v epohata na Makedonskata dinastija', in A. Grabar, *Izbrani sachinenija*, I (Sofia, 1982), pp. 112–31; O. Grabar and A. Grabar, 'L'Essor des arts inspirés par le cours princiers à la fin du première millenaire: princes musulmans et princes chrétiens', in *L'Occidente e l'Islam nell'alto medioevo*, Spoleto, 2–8 aprile 1964 II (Spoleto, 1965), 858–63 (Settimane di Studio del centro italiano di studi sull'alto medioevo, XII).

15   V. Beshevliev, *op. cit.*

16   *Ibid.*, pp. 142–4, 146–7, nos 19–21, 22, 26, 28, 29.

17   *Ibid.*, pp. 139–140, no. 15.

18   *Ibid.*, pp. 170–2, no. 46.

19   *Ibid.*, p. 144, no. 22.

20   Its date is generally assumed to be seventh–eighth century. See *Madarskijat konnik* (Sofia, 1956); R. Rashev, 'Konnikat v starobalgarskoto izkustvo', *Arheo-*

logija, 2–3, 1984, 60–71; Rashev is of the opinion that the representation of the horseman is a generalised image of the hero-ruler but not a personified hero, nor historical person or deity. He also stresses the 'epic intonation' in the inscriptions around the relief that link the relief with the culture of the ancient Turks practising the cult of the horseman.

21 *Madara. Razkopki i prouchvanija*, I–II (Sofia, 1934–6); S. Vaklinov, *op. cit.*, p. 172.

22 Zh. Aladzhov, 'Za kulta kam Tangra v srednovekovna Balgaria', *Arheologija*, 1–2, 1983, p. 82.

23 E. Moussakova, *op. cit.*, pp. 75–6.

24 *Ibid.*, p. 76.

25 O. Minaeva, *op. cit.*, p. 110.

26 B. Angelov, *Iz starata balgarska, ruska I sarbska literatura. III* (Sofia, 1978), pp. 69–98.

27 N. Dragova, 'Starobalgarsko voinsko skazanie ot IX vek', paper read at the 3rd international conference on the problems of folklore, Michajlovgrad, 1985.

28 *Ibid.*

29 S. Maslev, 'Pustinnozhitelska Madara', *Arheologija*, 1959, 3–4, 24–34.

30 I. Dujchev, 'Ezicheska i hristijanska Madara', in *Madara*. III (Shumen, 1992), 5–10.

31 A. Grabar, *op. cit.*, pp. 34, 170–1; *idem*, *L'Iconoclasme byzantin* (Paris, 1984), pp. 35–40.

32 D. Duchev, *Otgovorite na papa Nicolai i po dopitvanijata na balgarite/Responsa Nikolai papae ad consulta Bulgarorum* (Sofia, 1922), §33.

33 F. Kardini, *Istoki srednevekovogo ricarstva* (Moskva, 1977) (abridged trans. of F. Cardini, *Alle radici della cavalleria medievale*, 1982).

34 See D. Talbot-Rice, 'The Leaved Cross', *Byzantinoslavica*, II, 1, 1950: 72–81.

35 *Ibid.*

36 These are still unpublished. I express my gratitude to Dr Kazimir Popkonstantinov who showed me many of the finds in 1984.

37 The 'cross patoncée' is a cross whose arms have endings resembling toes, hence 'paton' (Fr.), A. Grabar, *L'iconoclasme*, p. 35. It represented the triumphal cross erected in Constantinople, A. Grabar, *L'Empereur*, p. 34. The same cross appears many times on vessels and objects of some Syrian treasures discovered near Apamea (Stuma, Riha and Hama) and dated sixth–seventh centuries which lead to the conclusion that a relic of the 'True Cross' was venerated there, M. Mango, *Silver from Early Byzantium. The Kaper Koraon and Related Treasures* (Baltimore, 1986), p. 77.

38 V. Beshevliev, 'Znachenieto na parvobalgarskija znak "Y"', *Izvestija na Narodnija muzej vav Varna*, XV (XXX), 1979, 17–23.

39 *Ibid.*; Aladzhov, *op. cit.*

40 R. Rashev, 'Za hronologijata I proizhoda na znaka "upsilon s dve hasti"', in *Prinosi kam balgarskata arheologija. I* (Sofia, 1992), pp. 96–102.

41 T. Totev, 'Edin bronzov medaljon sas znaci ot Preslav', *Arheologija*, 2, 1967, pp. 36–8.

42 R. Kostova, 'Za biblejskija smisal na edin rannosrednovekoven simvol', in *Balgarite v Severnoto Prichernomorie. Izsledvanija i materiali*. III (Veliko Tarnovo, 1994), 81–99.

43 D. Ovcharov, 'Za sadarzhanieto na edin balgarski rannosrednovekoven sim-vol', in *Balgarsko srednovekovie. Balgaro-savetski sbornik v chest na 70–godishni-nata na akad.* Ivan Dujchev (Sofia, 1980), pp. 294–8.
44 Vat. gr. 699, *The Christian Topography of Cosmas Indicopleustes.*
45 E. Moussakova, *op. cit.*, p. 118; see R. Kostova, *op. cit.*, fig. 8 which is not discussed by the author.
46 R. Kostova, *op. cit.*

4
Dualist Heresy in the Latin Empire of Constantinople

Bernard Hamilton

The Latin Empire of Constantinople came into being as a result of the diversion of the Fourth Crusade in 1204. At its greatest extent it comprised Bithynia, Thrace, the area south of the Marica including Philippopolis, and most of mainland Greece except for Epirus. Most of the Aegean and Ionian islands came under Venetian rule. Although the Empire came to an end in 1261, parts of the Peloponnese and of Attica remained in Latin hands until the Ottoman conquest, and some of the islands were ruled by Venice for much longer.

The presence of Christian dualists, Paulicians and Bogomils, in those territories is amply documented in the twelfth century, but after the reign of Alexius I it was the Bogomils who caused the imperial authorities most anxiety. The Paulicians were a readily definable group, who made no attempt either to conceal their beliefs, or in the Comnenian period, to make converts and were not therefore considered a great threat to the established Church.[1] When the Fourth Crusade captured Philippopolis in 1205 they found Paulicians there, and they subsequently burned down their quarter; but this was not because the Paulicians were heretics, but because they allied with Tsar Kalojan of Bulgaria against their new western rulers. This is the only mention of Paulicians in either the Latin or the Greek sources during the period with which I am dealing. So the rest of this chapter will be concerned with Bogomils.[2]

The Bogomils were particularly feared by the Byzantine authorities because their initiates were difficult to distinguish from Orthodox monks, and they were therefore uniquely well placed to infiltrate the Church of God.[3] Moreover, they were anxious to proselytise and during the twelfth century (and perhaps earlier) had sent missions to the West which met with great success in some areas, notably southern France and northern Italy. Their western followers were known as Cathars, and until at least 1200 they maintained close links with their mother churches in the Byzantine world. Indeed, some of our best information about the organisation of twelfth-century Bogomils is found in western sources. It is from them that we first learn that the Bogomils had adopted a system of episcopal government and that a schism had developed between the original moderate dualist Bogomils of Bulgaria and the absolute dualist Bogomils of Dragovitia, the area to the south of Philippopolis, who interpreted the Christian faith in terms of belief in two coeternal gods, or principles, of Good and Evil. By c.1170 the absolute dualist wing had gained control of the Bogomil Church of Constantinople, whose leader, Nicetas, visited western Europe where he presided at a Council in southern France and inculcated his version of the Bogomil faith among the Cathar leadership.[4]

From a combination of western and Byzantine sources we know that in the lands which came to make up the Latin Empire there had in the twelfth century been Bogomil communities in Constantinople, Dragovitia, Thessaly, Hellas, and among the Milingui, a Slav-speaking tribe who lived in the Taygettus mountains of the Peloponnese.[5] The western Church was alarmed by the rapid growth of Catharism, and was in a general sense aware that it had links with dualist movements in the Byzantine world, for that had been admitted by the first Cathar leaders brought to trial in a western court, those at Cologne in 1143.[6] Pope Innocent III, in whose pontificate the Latin Empire was conquered, was dedicated to the fight against dualism in East and West: in 1203, using the threat of coercion by the King of Hungary, he forced the leaders of the Bogomil movement in Bosnia to make their submission to the Catholic Church;[7] in 1209 he launched the Albigensian Crusade against the Catholic rulers of southern France who tolerated Cathars in their lands; and he may also have been responsible for the promulgation of anti-Bogomil legislation by the Bulgarian Tsar Boril in 1211 at the Council of Trnovo.[8] Yet Innocent took no action of any kind against the Bogomils in the Latin Empire of Constantinople.

This suggests that the Pope was not aware that there were any Bogomils in the Greek-speaking Byzantine lands. This is certainly possible because most of the evidence on which our knowledge of twelfth-century Byzantine Bogomilism is based would not have been available to him. He would not have known the Greek texts; while the Latin sources date mostly from after the foundation of the Inquisition in c.1229, when a good deal of information about the history of the Cathar churches and their links with the Bogomil churches was made available both by Cathar initiates converted to Catholicism, and through the depositions made to inquisitors by ordinary Cathar believers,[9] while some Cathar texts, such as the account of the Council of St Félix, also came into Catholic hands. The one text which Innocent could have read was the Contra Patherenos written in c.1170 by Hugo Eteriano, the Pisan lay adviser on western Church affairs to the Emperor Manuel Comnenus. Hugo wrote it in order to persuade the Emperor to take action against the Bogomils in Constantinople, and it is directed at a Greek audience, although the surviving texts are in Latin.[10] How far it was diffused in the West by 1200 is a matter which my wife and I are trying to resolve, as we are preparing an edition of this work. Should any reader have any relevant information about this text I should be glad to know about it.

But unless Innocent III had read Hugo's work, and there is no evidence that he had, he would have associated the Bogomils with Bulgaria, which was indeed the cradle of that faith, because it was by that name that their adherents were popularly known in western Europe.[11] He was aware that they had spread into Dalmatia and Bosnia, because these regions formed part of the Latin Church, and also because Prince Vukan of Dioclea had denounced the Ban of Bosnia to Innocent as a notorious protector of heretics.[12] But Innocent does not seem to have been aware that there were Bogomils in the Greek-speaking Byzantine lands. In this way the anomaly arose that at a time when the Pope was campaigning against heresy in the West and in the Balkans, the Bogomils of the Latin Empire were safe from any kind of prosecution.

There is no doubt at all that they were stlll there. The Orthodox Patriarch of Constantinople, Germanus II (1222–40), living in exile at Nicaea, not only preached against those in the Empire of Nicaea, but also sent an encyclical letter to the Orthodox clergy in the city of Constantinople, which he ordered to be publicly read on Sundays and holy days, putting the faithful on their guard against the Bogomils in the capital.[13]

Papal failure to take action against them was not an indication of any lack of zeal against dualism. The Albigensian Crusade finally came to an end in 1229 and was a victory for the forces of orthodoxy, those of the Pope and the King of France, and soon after that the papal Inquisition was established in parts of western Europe to enforce the heresy laws.[14] Pope Gregory IX (1227–41) tried to use parallel methods against the Balkan Bogomils; in 1235 unleashing a crusade led by the King of Hungary against Bosnia, whose rulers were accused of favouring dualists, and three years later attempting to use the same measures against Bulgaria, which had renounced its links with Rome and whose rulers were likewise suspected of tolerating Bogomils.[15] Both these initiatives were brought to an end by the great Mongol raid on Eastern and Central Europe in 1240–2.[16] But no action was taken against the dualists in the Latin Empire.

Nevertheless, more accurate information was being gathered about them by the Inquisition in Lombardy. Raynier Sacconi, a Cathar minister of some seventeen years' standing, was converted to Catholicism by the Dominican St Peter Martyr in c.1245 and later became chief Inquisitor for Lombardy.[17] In c.1250 he wrote a compendium about Catharism for the use of his assistants which contains up-to-date information also about the Bogomil churches. He lists the churches of the Cathars 'beyond the seas': 'The Church of Sclavonia. The Church of the Latins of Constantinople. The Church of the Greeks in the same city. The Church of Philadelphia in Romania. The Church of Bulgaria. The Church of Duguuithia. They all took their origin from the two last named.'[18]

The Church of Sclavonia was the Church of Bosnia, and it together with that of Bulgaria were, as we have seen, already known to the Catholic authorities. The Church of Philadelphia in Romania presumably referred to the Bogomils in the Byzantine Empire of Nicaea about whom the Patriarch Germanus had complained, but which was outside the political control of the Latin Emperor.[19] The Church of Duguuithia was an attempt to write Dragovitia (a name which caused all western authors great problems) and it, together with the Latin and Greek Churches of Constantinople were situated in territory claimed by the Latin Empire. Information of this kind continued to be built up by the inquisitors about the Bogomils in the East. Rainier's successor in Lombardy, Anselm of Alessandria, was particularly interested in Cathar history, and investigated the links between the Italian Cathars of his own day and the dualist churches of the Balkans and Greece.[20]

But this information about the Bogomil churches in the Latin Empire in some cases came too late for the Catholic authorities to act upon, even had they wished to do so. The area of Dragovitia, south of Philippopolia, had passed under Bulgarian rule again in c.1207. The Latin Kingdom of Thessalonica, where Bogomils had been active in the twelfth century, was for the most part conquered by the Despots of Epirus in 1224.[21] The Church of Milinguia in the Peloponnese, if it continued to exist in the thirteenth century, was effectively outside the jurisdiction of the Frankish Princes of Achaia. The Chronicle of the Morea relates that in c.1250 the chieftains of the Milingui made a treaty with William of Villehardouin in which they stipulated that: 'they would never... perform corvées nor pay levies...; but would give homage and service at arms, as they likewise had given to the [Byzantine] Emperor'.[22]

But at the time when Sacconi wrote the Duchy of Athens and the city of Constantinople were still in Frankish control, yet no measures were taken against the Bogomils there. Part of the reason for this may have been that in outward appearance the Bogomil leaders looked like Orthodox monks: the Orthodox clergy had from the beginning complained about this.[23] The Latin hierarchy would not therefore have been able to detect the Bogomil initiates unless they had questioned them about their beliefs. But the Inquisition was not established in any of the lands under Latin rule in the eastern Mediterranean during the thirteenth century, even though in 1267 Pope Clement IV had ordered that it should be, for his death a few months later and the long interregnum which followed meant that this command was never implemented.[24] Consequently there were no Latin clergy there whose duty it was to seek out heretics. But although this would explain why the Latin hierarchy took no action against Bogomils in the Duchy of Athens (if there were still any there) or in the city of Constantinople where they are known to have existed, it fails to explain why no action was taken about the Church of the Latins in Constantinople reported by Raynier Sacconi, whose members must have been western Cathars. I have argued elsewhere that there were Latin converts to Bogomilism who lived in Constantinople from c.1100, and that they played a central role in spreading dualist ideas in western Europe in the twelfth century, but there is no evidence that they were separately organised at that time.[25] Rainier Sacconi's report is the first record we have that there were two dualist churches in the capital in the thirteenth century. One possible explanation of this is as follows: in c.1170 the Bogomils of Constantinople had been absolute dualists of the Dragovitian *ordo*, but by c.1235, as the Patriarch Germanus's encyclical makes plain, the Greek Bogomils there had reverted to the moderate dualism of the Bulgarian *ordo*.[26] It is possible that at the time when this change occurred the Latin dualists in the city had remained true to their Dragovitian tradition and had come to form a separate church. This is, of course, only hypothetical.

The Cathar community in Constantinople under Latin rule was not negligible: Sacconi reports: 'The Church of the Latins in Constantinople has about fifty members.'[27] This does not sound very impressive, but two factors must be considered. First, he was trying to allay Catholic fears of Catharism by playing down the number of Cathars throughout the world; secondly, this number relates only to initiates. If one regards them as the ordained clergy of the Cathar Church, then their adherents should be estimated with a multiplier of about 100, giving some 5000 Latin Cathar believers in Constantinople in the 1250s.

Ironically, this large number of believers was partly caused by the work of the Inquisition in southern France. Only convicted and unrepentant heretics could be handed over to the secular arm for burning at the stake, which in practice meant that this punishment was confined to fully initiated Cathars. Cathar believers invariably recanted and had to perform penances before they were reconciled to the Catholic Church. In later times the more serious offenders among them were given prison sentences as their penance; but in the early years there were no prisons to which they could be sent, because imprisonment had not been a normal punishment before that time in most western countries.[28] An alternative penance, which some inquisitors imposed on young male offenders, was to do military service for a set number of years in the Latin Empire of Constantinople. In 1241–2, for example, the inquisitor Pet Seila, working in Quercy, imposed that penance on 93 men, which represented 15 per cent of all the sentences he gave in that year.[29] This provided the hard-pressed Latin Empire with free fighting men, but since by definition such men had Cathar sympathies they would have been natural supporters of the Cathar Church of the Latins in Constantinople.

In 1261, Michael VIII recaptured Constantinople and the Latin Empire technically came to an end. Bogomils remained there, but they must have been Greeks:[30] what happened to the Church of the Latins is not known, although there were many lands in Greece and the Aegean which remained in western political control and where the Inquisition did not operate to which its members could have gone. But in default of further evidence it is idle to speculate about this.

My initial reaction when I realised that the existence of the Church of the Latins in Constantinople was known to the Inquisitor of Lombardy at a time when the city was under western rule, yet no action was taken against it, was one of incredulity. It is true that the government of the last Latin Emperor, Baldwin II, was very weak and that he no doubt feared to stir up dissent among his Latin subjects,[31] but that the Papacy should not have tried to intervene, even if ineffectively, is astonishing. Yet it was so: there is not a single letter in the papal registers, which are complete from 1216 onwards, relating to the prosecution of dualists in the Latin Empire of Constantinople.[32] It was not until 1318 that Pope John XXII granted a faculty to the Dominican Order in Greece to appoint Inquisitors, and it is not known to me whether this permission was ever used.[33]

But I think that there is a simple explanation of papal inaction: Rainier Sacconi knew that there was a Cathar Church among the Latins of Constantinople but the papal curia did not.[34] The papal Inquisition in the thirteenth century was a very unbureaucratic institution: each inquisitor was personally responsible to the Pope, but there was no central committee to coordinate the work of individual inquisitors analogous to the sixteenth-century Holy Office.[35] The consequence of this was fortunate for some Cathars: the Latin Empire of Constantinople, as long as it remained under western rule, was a place where they might practise their faith in freedom. Indeed, some of them had even been sent there by the Inquisition as a penance for their support of Catharism in western Europe.

Notes

1 '[The Paulicians'] teaching is very like that of these blasphemers [i.e. the Bogomils], but their heresy is obvious and cannot harm anyone except those who hold it as inherited tradition; no-one is grieved or upset on their account.' Euthymius of the Periblepton, *Letter*, ed. G. Ficker, *Die Phundagiagiten: ein Beitrag zur Ketzergeschichte des byzantinischen Mittelalters* (Leipzig, 1908), p. 63, lines 14–16.

2 Geoffrey of Villehardouin, *La conquête de Constantinople*, *c.399*, ed. E. Faral, 2 vols (Paris, 1938–9), II, p. 210.

3 Anna Comnena, *Alexiad*, XV, viii, 1, ed. with French tr. B. Leib, 3 vols (Paris, 1937–45), III, p. 219.

4 D. Obolensky, 'Papa Nicetas: a Byzantine dualist in the land of the Cathars', *Harvard Ukrainian Studies*, 7 (1983), pp. 489–500; B. Hamilton, 'The Cathar Council of Saint-Félix Reconsidered', *Archivum Fratrum Praedicatorum* [henceforth *AFP*], 48 (1978), pp. 23–53.

5 The communities of Thessaly and Hellas are attested in the *Dialogue of Demons*, once attributed to Michael Psellus but more probably the work of Nicholas of Methone, M. Angold, *Church and Society in Byzantium under the Comneni, 1081–1261* (Cambridge, 1995), pp. 496–8. Y. Dossat, 'A propos du concile cathare de saint-Félix: Les Milingues', *Cahiers de Fanjeaux*, 3 (1968), pp. 201–14, was the first scholar to identify the 'Ecclesia Melenguiae' of the Saint-Félix document with the Milingui, but took this as additional evidence that the source was a forgery, because he did not accept that the Milingui could have been Bogomils. He gave no reason for this view, and there is none.

6 They claimed that '... hanc haeresim usque ad haec tempora occultatam fuisse a temporibus martyrum, et permansisse in Graecia et quibusdam aliis terris', reported Eberwin of Steinfeld, who conducted their trial, Appendix to the Letters of St Bernard, no. CDXXXII, in J. P. Migne, *Patrologia Latina* [henceforth *PL*], 182, col. 679.

7 *Pontificia Commissio ad redigendum codicem iuris canonici Orientalis*, series III [henceforth *CICO*, III], vol. II, *Acta Innocentii papae III (1198–1216)*, no. 36, ed. T. Haluscynski (Vatican City, 1944), pp. 235–7. Whether there was widespread Bogomilism in Bosnia during the later Middle Ages is a controversial issue.

I find the evidence for the presence of Bogomil communities there in the thirteenth century convincing, although I would concede that the extent of their influence may have been exaggerated by hostile sources. The classic presentation of the alternative view is that of J. V. A. Fine, *The Bosnian Church, a New Interpretation: a Study of the Bosnian Church and its Place in State and Society from the Thirteenth to the Fifteenth Centuries* (New York and London, 1975).

8   M. J. Popruzenko, 'Sinodik carja Borila', *Bulgarski starini*, 8 (Sofia, 1928); Y. Stoyanov, *The Hidden Tradition in Europe* (London, 1994), pp. 172–3.

9   The one possible exception to this is the *De Heresi Catharorum in Lombardia*, the work of an anonymous Cathar convert to Catholicism written before 1214, ed. A. Dondaine, 'La hiérarchie cathare en Italie, I. Le *De heresi Catharorum in Lombardia*', *AFP*, XIX (1949), pp. 280–312. 'Believers' / 'credentes' was the term used by the Catholic authorities to describe people who had Cathar sympathies and attended Cathar meetings, but who had not been received into the Cathar Church.

10 Two manuscripts are known to us: Seville, Bibliotheca Colombina, MS. 5.1.24, ff.67r–75v; Bodleian Library, Oxford, MS. Canon Pat. Lat., 1. Cf. A. Dondaine, 'Hugues Ethérien et Le'on Toscan', *Archives d'histoire doctrinale et litteraire du moyen âge*, 19 (1952), 67–134 (especially, pp. 109–14).

11 J. Duvernoy, *La religion des Cathares* (Toulouse, 1976), pp. 309–11.

12 Innocent III, *Regesta*, II, no. clxxvi, *PL*, 214, 725–6.

13 Ficker (ed.), *Die Phundagiagiten*, pp. 115–25.

14 Probably the best brief treatment in English of this topic remains W. L. Wakefield, *Heresy, Crusade and Inquisition in Southern France, 1100–1250* (London, 1974).

15 Stoyanov, *Hidden Tradition*, pp. 185–6.

16 D. Morgan, *The Mongols* (Oxford, 1986), pp. 136–41.

17 F. Sanjek, 'Raynerius Sacconi. O.P. *Summa de Catharis*', *AFP*, 44 (1974), pp. 32–5.

18 *Ibid.*, p. 50.

19 As Stoyanov rightly points out, the Bogomils in the Nicene Empire may have given this name to their church because of its association with the Revelation of St John the Divine, I, v.ll, *Hidden Tradition*, p. 164.

20 'La hiérarchie cathare en Italie, II: Le Tractatus de Hereticis d'Anselme d'Alexandrie, OP', ed. A. Dondaine, *AFP*, XX (1950), pp. 234–324.

21 The lordship of Boudonitza remained in Frankish hands for much longer. P. W. Lock, *The Franks in the Aegean 1204–1500* (London, 1995), pp. 60–2.

22 E. Lurier, tr., *Crusaders as Conquerors: the Chronicle of Morea* (New York, 1964), p. 160; cf. Constantine Porphyrogenitus, *De Administrando Imperio*, c.50, ed. G. Moravcsik, English tr. by R. Jenkins (Washington, 1967), pp. 232–5.

23 '. . . [the Bogomils] play-act the monastic and priestly way of life', Euthymius of the Periblepton 'Letter', ed. Ficker, *Die Phundagiagiten*, p. 63, lines 23–5; 'It appears that [Bogomilism in Constantinople] existed even before my father's time, but unrecognized; the sect of the Bogomils is very skilful at counterfeiting virtue . . . the evil is hidden under a cloak or a cowl.' Anna Comnena, *Alexiad*, XV, viii, 1, ed. Leib, III, p. 219.

24 'Inquisitores haereticae pravitatis etiam in provinciis orientalibus deputarunt
    Romani Pontifices', *CICO*, III, vol. V(I), *Acta Urbani IV, Clementis IV, Gregorii X*,
    ed. A. L. Tautu (Vatican City, 1953), pp. 73–4, No. 27.
25 'Wisdom from the East', in. P. Biller and A. Hudson (eds), *Heresy and Literacy,
    1000–1530* (Cambridge, 1994), p. 38.
26 Germanus relates that the Bogomils of Constantinople claim that the Devil is
    'the maker of all the visible creation and its king. Then they name him the
    son of God and brother of Christ...' Encyclical Letter, ed. Ficker, *Die Phun-
    dagiagiten*, p. 117, lines 27–8.
27 Sacconi, 'Summa', *AFP*, 44 (1974), p. 50.
28 B. Hamilton, *The Medieval Inquisition* (London 1981), pp. 30–59.
29 E. Albe, 'L'Hérésie albigeoise et l'Inquisition en Quercy', *Revue d'histoire de
    l'Eglise de France*, 5 (1910), pp. 280–91. This occurred despite the ruling of
    Gregory I in 1237 that those converted from heresy should not be sent 'ad
    partes transmarinas' to perform their penances, *CICO*, III, vol. III, *Acta Hon-
    orii III et Gregorii IX*, no. 220, ed. A. L. Tautu (Vatican City, 1950), pp. 296–7.
    This seems to have been understood as a veto on their going to the Holy
    Land.
30 Nicephorus Gregoras, *Byzantina Historia*, ed L. Schopen and I. Becker. *Corpus
    Scriptorum Historiae Byzantinae*, 3 vols (Bonn, 1829–55), II, pp. 728–30.
31 Lock, *Franks in the Aegean*, pp. 62–6.
32 The possible exception is the letter of Gregory IX, dated 7 December 1232,
    about a certain Theodora of Constantinople, who had separated from her
    husband on account of his 'spiritual fornication' described as 'haeretica
    pravitas'. Gregory ordered her husband to pay alimony, but did not specify
    the nature of his heresy. *CICO*, III, vol. III, no. 182, ed. Tautu, p. 251.
33 In 1318 John XXII granted a faculty to the Prior Provincial of the Dominicans
    in Greece to appoint Inquisitors where needed. The Order had houses in
    Crete, Negroponte, Thebes, Clarenza and Modon. It is not known to me
    whether this faculty was ever used. *CICO*, III, vol. VII (II), *Acta Iohannis
    XXII*, no. 18, ed. A. L. Tautu (Vatican City, 1952), pp. 30–3.
34 I do not know when the Holy See obtained a copy of Sacconi's work: the
    oldest extant MS, Vat. Reg. 428, ff. 3r–8v, dates from c.1260–70, and once
    belonged to Queen Christina of Sweden, but its earlier provenance is
    unknown to me. The other copies which are now in the Vatican Library
    date from the fourteenth century or later. Sanjek, 'Raynerius Sacconi', *AFP*,
    44 (1974), pp. 38–41.
35 'Inquisition', in F. L. Cross and E. A. Livingstone (eds.), *The Oxford Dictionary
    of the Christian Church* (3rd edn, Oxford, 1997), pp. 836–7.

5
The Religious Situation of the Hungarian Kingdom in the Thirteenth and Fourteenth Centuries

Ioan-Aurel Pop

Introduction

From both a denominational and a cultural point of view, the Western, Roman-German and the Eastern, Roman-Byzantine (and Slavic-Byzantine) traditions were in evidence in the territory of Hungary during the first two centuries of Arpad Kingdom (approx. 1000–1200). Pre-Hungarian Pannonia had witnessed both models of civilisation and, at the time of the Hungarian invasion, the region was inhabited by Christians.[1] The first pagan Hungarian chieftains to come into direct contact with the Christian faith, namely Bulcsu and Gylas (Gyula), around the middle of the tenth century, are said to have received the baptism in the Byzantine rite of Constantinople. Gylas brought with him to his country Bishop Hierotheus, consecrated by Patriarch Theophylactus. The location of the new Episcopal residence is not known, but it may have been somewhere in the south-eastern or eastern part of the then Hungarian-dominated territory.[2] This fact did not greatly influence the Hungarian population, which remained largely pagan, but it did increase Byzantine influence in the area. Even after the mass conversion of the Hungarians, after the year 1000,[3] the institutions of the Eastern Church remained and developed in the new kingdom. Alongside Judaism, Islam, several 'heresies' and some so-called 'heathen' cults, the two Christian denominations co-existed peacefully in the Hungarian kingdom until around the year 1200. As new countries and provinces came to be included (most often through military conquests) and various populations (from the West or the East) were colonised and settled, the ethnic and confessional mosaic of the Hungarian kingdom became diversified, encouraged – in a seemingly successful formula – on his death-bed by the founding king, (Saint) Stephen I: *Regnum unius linguae fragile et imbecille est.*[4] Even if the authenticity of this 'last will' of Stephen is yet to be determined, the message was reflected in the multi-ethnic and pluri-confessional character of the medieval Hungarian state.

Yet, at the beginning of the thirteenth century, the relative harmony which gave strength to the diversified society of Hungary changed abruptly from an ethno-confessional point of view, disturbing, often quite severely, the entire development of the medieval state. Starting with Innocent III (1198–1216), the Fourth Crusade (1202–4) and the Fourth Lateran Council (1215), the attitude of the Roman Church towards the members of the Eastern Church became quite radical.[5] This, in turn, led to a harsher policy of the 'apostolic kingdom' of Hungary towards 'schismatics' and non-Christians. The key moments of this policy, in the thirteenth and fourteenth centuries, offer an opportunity for an assessment of the proportions of the various denominations in Hungary. These are the following: 1. the period between the fall of Constantinople under the 'Latins' (1204) and the great Mongolian invasion (1241); 2. the period of the second half of the thirteenth century, namely the end of Bela IV's reign (1235–70) and the reign of Ladislas IV the Cumanian (1272–90); 3. the Angevin kings of Hungary, with a less important sequence under Charles Robert (1308–42) and with another one, under Louis I (1342–82), which marked the most important effort to create a Catholic unity of entire peoples, denominations and religions, from Hungary and neighbouring areas.

The policy of Innocent III and its consequences for the Hungarian kingdom

Innocent III[6] saw the Pope, vicarius Christi, as head of the universal Church (universalis ecclesia) which included the entire body of Christ, in heaven and on earth, Church and Christians, priesthood and kingship. In other words, the Holy Father had *spiritualium plenitudinem and latitudinem temporalium*, both captured in the concept of *plenitudo potestas* or 'complete power' over the Church and the 'Christian people'. Therefore, in the eyes of the Pope, who co-ordinated the action, the conquest of Constantinople and the creation on the ruins of Byzantium of the Eastern Latin Empire were political as well as religious accomplishments. On the one hand, it ended the 'usurpation' of the Roman Empire by the Byzantines, or Greeks, and on the other it annihilated the

'Greek schism'. In a letter dated 13 November 1204, the Pope declared
that the passing of the Empire of Constantinople 'from the Greeks to the
Latins', 'from the disobedient to the devout', 'from the schismatics to
the Catholics' etc. was a divine decision.[7] Therefore, 'schismatics' everywhere had tacitly to accept the Roman faith. Those who refused were to
share the fate of the Greeks who, because of their 'rebellion' and 'disobedience' to Rome, 'had been given in prey and pillage' (*dati fuerint in
direptionem et praedam*).[8] But confiscation of assets or looting was one of
the punishments which, according to the canons, were to be inflicted
on the heretics.[9] Thus, the Pope tended to identify schism and heresy
and apply the confiscation of assets to their adherents.[10] The theoretical
argument supporting the attitude of the Holy See – increasingly inflex-
ible in this matter – was the question of the Holy Ghost as originating
from the Son as well as the Father (filioque). Innocent III suggested, and
those who followed him confirmed, that, by rejecting the filioque,
namely by making a dogmatic 'mistake', the Orthodox had gone
beyond the lesser 'fault' of schism (= breaking ties with Rome), becoming guilty of the greater 'fault' of heresy.[11] Thus, the 'schismatic' owners
of assets (especially lands) – whether sovereigns or mere feudal lords –
became *iniusti possessores* and were liable to dispossession, by means of a
Crusade if need be, after the formula developed post factum and applied
to the Byzantine Empire.[12]

The fall of Constantinople into other, but still Christian, hands came
as a shock for the élites of Europe. At the beginning, the Pope himself
seemed surprised by the abuse committed there, but quickly overcame
all petty scruples and hastened to give a doctrinaire justification to
this act, to absolve the guilty and find them merit in the creation of
Christian unity. Then the initial enthusiasm diminished, as parts of
the Byzantine world began to recover and the Orthodox peoples of the
Balkans, from Eastern and South-Eastern Europe, refused to abandon
their doctrine. But 'apostolic kingdoms', such as the Hungarian one,
adopted, from the arsenal of the papacy, through the monastic orders, if
not a carefully argued theoretical basis, then at least a practical way of
punishing the disobedient 'schismatic', by deeming them unworthy of
land ownership and its privileges.

The main point of attention were the 'Greek', Eastern bishoprics and
monasteries, which had to pass into the hands of the 'Latin' hierarchies,
or to the Western monastic orders.[13] It is estimated that, by 1241,
approximately 600 Orthodox ('Greek') monasteries had been founded
in Hungary, of which only 400 have been located.[14] The impossibility of
finding the remaining 200 results from two causes: on the one hand,

after 1204, many 'Greek' monasteries passed into the hands of the Western friar orders, and on the other, the regions where there were large numbers of these monasteries (Banat, Crişana, Maramureş, Transylvania etc.) have been less well documented than other areas.[15] For the same period (before 1241), we know of fewer than 200 monasteries of the Western monastic orders in Hungary.[16] The 600 monasteries of the Byzantine rite known before 1241 illustrate the substantial number of Eastern parishes, bishoprics and, above all, believers. So many monasteries could only function where there was a numerous Orthodox population. Even if we were to accept that in Transylvania, Banat, Crişana and Maramureş only 60 of these 600 monasteries really existed (as compared to the 25–30 Catholic – or converted into Catholic – ones known at that time),[17] this number would also reveal a considerable Orthodox presence in the eastern part of the kingdom. This presence is also shown by the important measures taken against the 'schismatics' from Transylvania and neighbouring areas after 1204: the passing of the Greek monasteries in Crişana under the jurisdiction of a 'Latin' bishopric (16 April 1204);[18] the establishment of an Orthodox bishopric from 'the land of Knez Bela's sons', through the Kalocsa bishopric, under the jurisdiction of the apostolic see (3 May 1205);[19] in 1205–6, King Andrew II granted the Cârţa (Kerch) monastery, as foundation place, 'land taken from the Romanians' (*terram ... exemptam de Blaccis*),[20] between the Olt and the Carpathians, in southern Transylvania. This was a territory previously organised from a political and religious point of view by the Romanians and now taken out of their jurisdiction;[21] some time between 1204 and 1215, a Hungarian noble family had taken, with the king's help, 'from the hands of schismatic Romanians' (*de manibus Vallacorum schismaticorum*) the castle and district of Medieş (Megessalla), located in the Satu-Mare area, in northwestern Transylvania.[22] This measure marks a new stage in the papal initiatives, which turn from the subordination of bishoprics and monasteries to the denominational issue of the Eastern believers itself. In this respect, the Lateran Council of 1215 (Lateran IV) clearly stated the Pope's view of his relations with the Eastern denomination in terms of ritual, hierarchy, ecclesiastical obedience and discipline. On this basis, applying to 'schismatics' the procedure designed for heretics, they embarked upon a vast action of cancelling the 'schism', of circumscribing Western canons to the Orthodox world.[23] In 1231–5, swayed by the Pope, the Hungarian King Andrew II (in conflict with his son Bela) took harsh measures of persecuting and even annihilating the Jews (Mosaics), the Saracens and the Ishmaelites (Muslims) and the 'false Christians' (those not subject to the Roman Church).[24] In 1231, the archbishop of Esztergom denounced the danger – for Church and faith – of giving to the Saracens and the Jews the tasks of lease-holders of the royal revenues and said that the true believers were kept in an inferior position and so they were prone to apostasy.[25] That is why, also in 1231, King Andrew II was obliged to ban the access of Jews and Muslims to public services, and in 1233, during the assembly 'from the skirt of the Bereg forest', he imposed on them an obligation to wear distinctive marks to differentiate them from the Christians.[26] The authorities paid attention to the 'schismatics' as well. In 1234, the 'peoples' (*populi*) called Romanians of the Cumanian bishopric – politically and ecclesiastically organised in their own structures[27] – had several bishops of the 'Greek rite' and refused to obey the Roman bishop.[28] For the time being, the Pope was only recommending 'respect for the rite of each nation' and 'a Catholic bishop fit for that nation', but the recommendations were far from being respected in the Hungarian kingdom. Therefore, at least as far as the Romanians were concerned, the results of the proselytising effort were almost non-existent in this period. On the contrary, the 1234 papal letter regarding the aforementioned Romanians of the Cumanian bishopric showed that 'Hungarians', 'Teutons' and 'other true believers' in the Hungarian kingdom elected to live with these Romanians and, 'forming with them one people' (*populus unus facti cum eisdem Wallathis*), they ignored the Catholic prelate and obeyed the Byzantine bishops.[29] The Italian friar Rogerius, resident for a while in Oradea, mentioned among the causes which had destroyed the Hungarian capability to resist the Mongols the attempts of Bela IV to unify so heterogeneous a kingdom: 'Because, due to the many differences and various rites, almost the entire Hungarian kingdom had been tarnished and the king was doing his best to reform it...'.[30] Because of this serious situation, illustrated by the adherence of a large number of the inhabitants to denominations and religions other than the Catholic one – 'Hebrews' (Jews), 'Saracens' (Muslims) and 'false Christians' (Orthodox), are described as dominant in the country and attracting Catholics to their rites – and because of the inefficient measures taken to consolidate the Roman faith, an interdiction was laid upon the Hungarian kingdom (in 1232).[31]

Catholics and non-Catholics in the times of Ladislas IV the Cumanian

After the great Mongolian invasion (1241–2), the issue of the 'schism', both within Hungary and outside, like that of the expansion of the

Roman faith at any cost, were subordinated to the needs of defence against 'heathen' attack. Of course, later on, under the pretext of fighting the Tartar danger, the Hungarian kingdom was to proceed once again with expansionism in Eastern Europe. An example was the attempt to colonise the Hospitallers in Ţara Severinului (*terra de Zeurino*), in 1247, at the expense of some incipient Romanian states. But the policy of force adopted by Hungary ran out of resources and had to stop. Under kings like Bela IV (after 1260), Stephen II (1270–2) and Ladislas IV (1272–90), the crusader spirit was replaced by that of negotiations and agreements with the Tartars (abroad) and by an increased tolerance towards pagans and the 'schismatics' (at home). At the Council of Lyon (1274) – where, formally, 'the union' of the two Christian churches was accepted – the Hungarian kingdom was denounced as a place in which Cumanians (colonised in the centre of the country, during the reign of Bela IV, around 1238–9 and finally settled there after the Mongol invasion) dominated state politics and drew the inhabitants to the 'perversity of their rite', as a place in which 'heretics and schismatics were openly defended'.[32] The king himself was born of a Cumanian mother, had taken up Cumanian habits and, according to one source, had secretly received Orthodox baptism.[33] Consequently, the papal legate Philip, Bishop of Fermo, was sent to Hungary (Poland and Cumania) to restore the Catholic faith in the region, create a unified Catholic religion and bring the Arpadian kingdom back to the forefront of the crusade. To this end, in 1279 Philip summoned the synod of Buda, aiming to discourage the other cults in Hungary: 'The Jews, Saracens, Ishmaelites and other pagans' were to wear distinctive marks on their chests when outside their houses; 'schismatics' and 'any others outside the Catholic faith' were not allowed to take any public office; Eastern priests were prevented from serving and from building new churches or chapels, 'Christians' (Catholics) were not allowed to take part in the services of 'schismatics' or to receive the Eucharist from them; 'Christians' had to be kept away from 'schismatics' by means of ecclesiastical punishments and, if need be, through the intervention of the 'secular arm'.[34] Shortly after the synod, Pope Nicholas III asked King Ladislas to drive the 'schismatics' and heretics out of Hungary.[35] These measures had no practical consequences, as the king refused to comply with them. Thus, the Roman faith remained for a while of secondary importance in Hungary, and on the death of the sovereign the Pope began an investigation to discover whether Ladislas had died as a 'heretic', a 'schismatic' or a Catholic.[36]

Religious beliefs in Angevin Hungary (1308–82)

The Angevins initiated a tremendous enterprise of unifying the varied structures in the Hungarian kingdom, of imposing respect and consolidating the Western feudal order. An important role in this enterprise was played by the Catholic faith. It is impossible to determine the number of Catholics in Hungary at that time, but the record of papal taxes for 1332–7 could give an idea as to their presence in Transylvania, Banat and Partium, by correlating it with other sources.[37] This document shows that in the aforementioned territories there were 954 localities with Catholic parishes. By 1350, there were 2552 confirmed localities. We may therefore estimate that, by 1331–40, there were 2100–2200 localities in that area (some villages had disappeared in the meantime, but others were not included in the written documents). Consequently, the villages with Catholic parishes represented 43–45 per cent of all Transylvanian localities, and the Catholic population may have oscillated between 35 and 40 per cent of the entire population (we know for certain that in many villages with Catholic parishes one could find a population of Orthodox denomination). We can say that the documentary proof, and the percentages are relative, generally shows that in Transylvania, in the 1330s, non-Catholics represented almost two-thirds of the population.

Obviously, the ambitious Louis I d'Anjou (1342–82) could not be satisfied with such a situation. Led by the Holy See and helped by the Western monastic orders (for the most part Franciscans), the king initiated the greatest effort of Catholic proselytising ever seen in Hungary and neighbouring regions. Documents frequently mention 'the multitude of schismatics, Philistines (= heretics, Iasians), Cumanians, Tartars, heathens and non-believers' in or around the Hungarian kingdom, for whom Louis obtained the right to found churches, to compel them to accept Catholic baptism or be expelled; in 1364, Bosnia was dominated by 'countless heretics and Patarenes (= heretics)', against whom a 'crusade' was conducted; in 1356, there is reference to 'all inhabitants of Transylvania, Bosnia and Slavonia who are heretics' as well as to the 'heretics and schismatics' of Serbia.[38]

After 1360, the effort of accomplishing 'unity of faith' – Catholic, of course – in Hungary and in neighbouring regions gained momentum through the involvement of the Franciscan order alongside the king. The main targets were the Romanians, Serbs, Bosnians and Bulgarians of the kingdom, who were supported by the independent states of their fellow nationals on the Hungarian borders. They were accused of religious errors (refusal of the *filioque*, baptism, the Eucharist and the use of leavened bread and, above all, the denial of papal supremacy and of the universalism of the Roman church) as well as secular ones: evils committed against the 'Christians' (= Catholics) 'together with those beyond the borders of the kingdom who share their language and sect', disobedience to local Catholic lords, violent recovery of confiscated assets (labelled therefore as thieves, criminals, rebels etc.).[39] The solutions were conversion, banishment or even extermination. Sources show that all three were applied, but with no significant results. For instance, the Franciscan friars claimed that in one year (approx. 1380) they baptised 400 000 'schismatics' into the Roman rite, which is exaggerated, even if reduced ten times, but suggests the large number of Orthodox believers in Hungary.[40] In 1374, the Pope knew that part of 'the numerous Romanian nation' which lived 'on the borders of the Hungarian Kingdom' had agreed to abandon the 'Greek schism' but was still hesitating, while many Romanians absolutely refused to do so because they 'were not satisfied with the sermons of Hungarian priests' and demanded a prelate who could 'speak the language of their nation'.[41] Another solution was applied after 1366: King Louis allowed Transylvanian noblemen – at their request – to exterminate the Romanian malefactors and rebels and establish the Catholic faith as a mandatory prerequisite for all noblemen.[42] Finally, around 1363–4, some of the Romanian leaders from Maramureş, led by Voivode Bogdan, were forced to leave their country and cross the mountains into Moldavia, accelerating the formation of the east Carpathian Romanian state.

All these events reveal the king's intention of imposing the principle of an official religion (*religio recepta*), essential in state policy after the Reformation, especially in Transylvania – where the Catholic faith was clearly endangered and where Louis I resided for about six months in 1366.[43] The Orthodox, theoretically unfit for land ownership, could no longer have access to the exercise of power. Therefore, the Romanian élite in Transylvania never received global privileges (like the noblemen, the Saxons and the Szeklers) and they were excluded from the estates. But these actions were not enough to ensure the success of the royal plans to accomplish unity of faith in Hungary.

In the fifteenth century, the humanist Antonio Bonfini, praising Louis the Great, reviewed his great accomplishments at a denominational level and performed a survey of the religious policy of this Angevin king of Hungary. Thus, the king is praised for his unshaken belief and for the following deeds: the promise to make the Jews of Hungary 'become Hungarian' through baptism or to expel them in case of refusal (which eventually happened); his support of the monastic orders; his creation of churches and monasteries; the guidance of the corrupted Cumanians, of the Patarenes (= heretics) from Bosnia and of other 'perverts', of whom many returned to their sin; therefore, said Bonfini, 'according to general opinion, faith in Hungary was so considerably enlarged and increased that more than one third of the kingdom was penetrated by the holy law' (*praeter omnium opinionem, religio in Ungaria nimis amplificata, et usque adeo propagata, ut plus tertia regni parte in divinium usum possideret*).[44] Bonfini makes these statements one century after the events, based on written sources and other testimonies; he has no interest in playing down the number of Catholics in the kingdom – on the contrary – and the expression 'according to general opinion' shows that the ratio of over one-third Catholics in Hungary in the year 1380 seemed natural, albeit after a proselytising effort such as the one carried out by Louis I.

Conclusions

It is known that medieval society was characterised by ethnic, linguistic, confessional, political, economic, social and cultural peculiarities. The age of historiographic romanticism had left us a series of clichés, periodically revived by interests alien to scholarship, regarding the unity and uniformity of some medieval states. The Hungarian kingdom is such an example.

From an ethnic point of view, Hungary came to include several population strata: 1. ethnic elements found by the Hungarians on their invasion of Pannonia and partially driven into fringe or even extra-Pannonian areas: Slavs, slavicised Bulgarians, Romanians, Germanic peoples (Gepidae etc.), Byzantines, old groups of Avars, Khazars (maybe Szeklers) who had arrived in advance of the Hungarians, 'Latins'; 2. peoples and populations conquered through the Hungarian campaigns of the tenth–fourteenth centuries: Slovaks, Romanians from Banat, Crişana, Transylvania and Maramureş, Serbs from the south, new Bulgarians, Ruthenians etc.; 3. groups that arrived through migrations and colonisation, with military, economic, demographic, political purposes, some from the west – 'Latins', urban Germans, Flemings, Saxons – and others from the east – north Iranian, Khoresmians, Caucasian Alanians (Sarmatians), Bashkirs, Petchenegs, Udae, Cumanians, Jews, Iasians (also called 'Philistines' or 'Jazyges') etc.; 4. peoples and ethnic groups who came to Hungary through marriage alliances, dynastic unions, through agreements or combinations of diplomacy and military power (such as Croatians, Bosnians, Italians from Dalmatia). In fact, the government programme of the Arpadian dynasty was based, as we have seen, on ethno-linguistic pluralism. This led also to religious and denominational pluralism. In Pannonia, the Eastern Christian faith is older than the Western. Byzantium dominated the region of the Middle Danube until the tenth century, and the mission to Moravia of the monks Cyril and Methodius in the ninth century had strong echoes in pre-Hungarian Pannonia. Documented evidence of approximately 600 Orthodox monasteries in Hungary before 1241 reveals the importance of the Eastern rite. Its followers were mainly Romanian, Serb, Bulgarian and Ruthenian. Other non-Catholics were the Jews (Mosaics), the Petchenegs, the Udae, the Cumanians, new Bulgarian arrivals, the Hungaro-Bashkirs and other Asian groups, some of whom had fled from the Mongols, and remained pagan or adopted Islam. The conversion drive had little effect on these peoples until the fourteenth century. In addition, the population of the kingdom manifested certain heresies (Bogomils, Patarenes etc.) which, eventually, came also to include the 'schism'.

After 1204 we witness the beginning and the intermittent development of a levelling action meant to create 'Catholic unity', as the power of the Western church grew in Hungary. The action was co-ordinated by the Holy See itself, but actually carried out by the apostolic Hungarian state (king, Hungarian church), with the help of the Catholic monastic orders. Its overall failure was due to known causes, which include: the large number of non-Catholics in Hungary and neighbouring regions; intransigence towards the converts; the often too-harsh intervention of the 'temporal arm' (the secular political-military element); the Mongolian invasion and then the dominance in the area of the Golden Horde; the tendency of some Hungarian political forces (even the kings) temporarily to abandon the front-line of the crusade and co-operate with non-Catholics; the combination of conversion factors and political-social and ethno-linguistic ones, intended firmly to integrate those attracted into the Hungarian kingdom and to level its structures etc.

In such a situation, according to the known sources – which do not offer statistical data, but only estimations – we can say that until the second part of the fourteenth century, the proportion of non-Catholics in the Hungarian kingdom was greater than that of Catholics.

Notes

1 Gyula Moravcsik, *Byzantium and the Magyars* (Budapest, 1970), p. 104; Ioan-Aurel Pop, *Romanians and Hungarians from the 9th to the 14th Century. The Genesis of the Transylvanian Medieval State* (Cluj-Napoca, 1996), pp. 80–90.

2   Alexandru Madgearu, 'Misiunea episcopului Hierotheus. Contribuții la
    istoria Transilvaniei şi Ungariei in secolul al X-lea', *Revista istorică*, 5, 1994,
    nr. 1–2, pp. 147–54.
3   György Györffy, 'La Christianisation de la Hongrie', in *Harvard Ukrainian
    Studies*, 12–13, 1988–9, pp. 61–73.
4   Jenö Szücs, 'The Peoples of Medieval Hungary', in *Ethnicity and Society in
    Hungary*, ed. F. Glatz (Budapest, 1990), passim; Ioan-Aurel Pop, *The Ethno-
    Confessional Structure of Medieval Transylvania and Hungary* (Cluj-Napoca,
    1994), passim.
5   V. Iorgulescu, 'L'Eglise byzantine nord-danubiènne au début du XIII-ième
    siècle, Quelques témoignages documentaires aux alentours de la quatrième
    Croisade', *Byzantinische Forschungen. Internationale Zeitschrift für Byzantinistik*
    (Amsterdam), 22, 1996, pp. 53–77; Constantin I. Andreescu, 'Reacţiuni orto-
    doxe în contra catolicizării regiunilor carpato-dunărene in prima jumătate a
    sec. al XIII-lea', *Biserica Ortodoxă Română* (Bucharest), 1938, nr. 11–12,
    pp. 770–9.
6   Yves Congar, *L'Eglise. De Saint Augustin à l'époque moderne* (Paris, 1970),
    pp. 193–5; Şerban Turcuş, *Christianitas şi românii: note privind locul român-
    ilor în realitatea creştină medievală occidentală la începutul secolului XIII*,
    forthcoming.
7   *Acta Innocentii PP. III (1198–1216)*, ed. Th. Haluscynskyi (Vaticano, 1944),
    pp. 277–8; Şerban Papacostea, *Românii în secolul al XIII-lea. Între Cruciată şi
    Imperiul Mongol* (Bucharest, 1993), p. 50. See also P. L'Huillier, 'La nature des
    relations ecclésiastiques gréco-latines après la prise de Constantinople par les
    croisés', *Akten des XI. Internationalen Byzantinisten kongresses* (München,
    1958), pp. 314–20; W. de Vries, 'Innozenz III. und der chrisciliche Osten',
    Archivum Historiae Pontificiae, 3, 1965, pp. 87–126.
8   Ş. Papacostea, *op. cit.*, p. 54.
9   H. Maisonneuve, *Etudes sur les origines de l'Inquisition* (Paris, 1942), pp. 170–1.
10 Ş. Papacostea, *op. cit.*, p. 54.
11 *Ibid.*, p. 51.
12 *Ibid.*, p. 54.
13 The names 'Greek' and 'Latin' are not used here in an ethnic sense, but a
    denominational one, denoting respectively 'Eastern' (Orthodox) and
    'Roman' (Catholic). Consequently, when 'Greek monks' in Hungary are
    mentioned, these are Orthodox friars.
14 G. Moravcsik, *op. cit.*, p. 114.
15 For instance, in Banat we know of the transformation of a 'Greek' monastery
    into a Western one since the eleventh century (Ioan-A. Pop, *Romanians*, pp.
    129–40), and in the twelfth century we have two other cases in the same
    region, those of Bistra and Pâncota (Suzana Heitel, 'Monumente de artă din
    secolele X-XII din Banat', research presented on 8 April 1997 at the Roma-
    nian Academy, Bucharest). Recently, researcher Mircea Rusu has stated that
    of the approximately 20 Byzantine monasteries attested until 1250 in north-
    ern Banat and Partium, more than half were taken over by the Western
    Catholic orders, especially after 1204, due to the new papal policy (examples:
    Hodoş, Geled, Bistra-Bizere, Eperyes, Pordan, Kenez, Cubiş); see Mircea Rusu,
    'Biserici, mânăstiri şi episcopii ortodoxe', in *Istoria Românie. Transilvania*,
    vol. I, ed. Anton Drăgoescu (Cluj-Napoca, 1997), pp. 300–10. The forced

transformation of many 'Greek' monasteries into similar Catholic institutions, especially after 1204, makes almost impossible the precise estimation of the proportion occupied by these two kinds of monastic structures in thirteenth-century Hungary.

16 J. L. Csoka, *Geschichte des Benediktinischen Mönchtums in Ungarn* (Budapest, 1980); J. Török, *Szerzetes-és lovagrendek Magyarországon* (Budapest, 1990); *Korai magyar történeti lexikon (9–14 század)*, ed. Kristo Gyula, Engel Pál, Makk Ferenc (Budapest, 1994), pp. 96–7, 139–40, 171–2, 218–19, 558–9.

17 See also Paul Niedermaier, *Der mitteralterliche Stadtebau in Siebenbürgen, im Banat un im Kreischgebiet*, vol. I (Heidelberg, 1996), p. 51.

18 *Documente privind istoria României*, C. Transilvania, veac. XI–XII–XIII, vol. I, pp. 28, 367 (hereafter *DIR*).

19 G. Fejer, *Codex diplomaticus Hungariae ecclesiasticus ac civilis*, vol. II, pp. 459–60; *DIR*, C, veac. XI–XII–XIII, vol. I, p. 29.

20 Franz Zimmermann and Carl Werner, *Urkundenbuch zur Geschichte der Deutschen in Siebenbürgen*, vol. I, pp. 26–8; *DIR*, C, veac. XI–XII–XIII, vol. I, pp. 199–200, 379–80.

21 Dan N. Busuioc-von Hasselbach, 'Mănăstirea cisterciană Cârța în secolul al XIII-lea. Contribuții la istoria Țării Făgărașului în evul mediu timpuriu' (doctoral thesis) (Cluj-Napoca, 1997), pp. 46–58. Several authors have considered that there had previously been an Orthodox monastery on the land where the Cistercian monastery of Cârța was founded. See *ibid.*, pp. 79–80, and also note 430 (on p. 349).

22 Aloisie L. Tautu, *Litterae Gregorii papae XI in causa Valachorum de Megesalla (Medieș in Transilvania septemtrionali) destinatae* (Rome, 1966), passim; Francisc Pall, *Romanians of Transylvania in the Middle Ages* (Cluj-Napoca, 1993), passim; Ș. Papacostea, *op. cit.*, pp. 73–5.

23 Ș. Papacostea, *op. cit.*, p. 72.

24 *DIR*, C, veac. XI–XII–XIII, vol. I, pp. 271–2.

25 Maria Holban, *Din cronica relațiilor româno-ungare în secolele XIII–XIV* (Bucharest, 1981), p. 54.

26 Charles d'Eszlary, *Histoire des institutions publiques hongroises, I* (Paris, 1959), pp. 387–93.

27 S. Papacostea, *op. cit.*, p. 63.

28 *Documenta Romaniae Historica*, D. Relații între Țările Române, vol. I, p. 20.

29 *Ibid.*

30 Ș. Papacostea, *op. cit.*, p. 72.

31 A. Theiner, *Vetera monumenta historica Hungariam sacram illustrantia*, vol. I (Roma, 1859), pp. 107–11.

32 Ș. Papacostea, *op. cit.*, p. 162. See András Pálóczi Horváth, *Pechenegs, Cumans, Iasians. Steppe Peoples in Medieval Hungary* (Budapest, 1989), pp. 77–82.

33 Ovidiu Pecican, 'Roman și Vlahata. O gestă slavo-română scrisă in Bihor (sec. XIV)', in *Familia* (Oradea), series V, year 28, 1992, 5, passim.

34 *Antiquissimae constitutiones synodales Provinciae Gneznensis maxima ex parte nunc primml e codicibus manu scriptis typis mandatae*, ed. Romualdus Hube (Petropoli (Saint Petersburg), 1856), pp. 159–62.

35 Ș. Papacostea, *op. cit.*, p. 163.

36 *Ibid.*

37   *DIR*, C, veac. XIV, vol. III, pp. 41–253; Ştefan Pascu, *Voievodatul Transilvaniei*,
     vol. I (Cluj, 1972), pp. 231–2; vol. II (Cluj, 1979), pp. 25–35, 442, 446–7 etc.
38   Ioan-Aurel Pop, *Ethno-Confessional Structure*, pp. 28–32.
39   Dionysius Lasic, O. F. M., 'Fr. Bartholomaei de Alverna, Vicarii Bosnae,
     1367–1407, quaedam scripta hucusque inedita', in *Archivum Franciscanum
     Historicum*, 1, year 55, 1962, nr. 1–2, pp. 72–6; Ş. Papacostea, *Geneza statului
     în evul mediu românesc. Studii critice* (Cluj-Napoca, 1988), pp. 94–5.
40   Ş. Papacostea, *Geneza*, p. 91. The entire population of Transylvania (with
     Banat, Crişana and Maramureş) could have been, around 1350, 700 000–
     800 000 inhabitants, showing how far from reality the Franciscan data were
     in suggesting a figure of 400 000 converts in one year.
41   E. Hurmuzaki, *Documente privitoare la istoria românilor*, vol. I/2 (Buchurest,
     1890), p. 217.
42   Ş. Papacostea, *Geneza*, p. 91.
43   *Ibid.*, p. 88.
44   Antonii Bonfinii, *Rerum Ungaricarum decades quatuor cum dimidia* (Basileae
     (Basel), 1568), decadis II, liber X, p. 377.

6
Crypto-Christianity and Religious Amphibianism in the Ottoman Balkans: the Case of Kosovo

Noel Malcolm

Crypto-Christianity was a widespread phenomenon in the Ottoman Empire, but not a common one. A map of the Empire, with the crypto-Christian communities coloured in, would show specks and patches here and there, scattered over a very wide area; but the communities in each case were fairly small ones, leading, of necessity, somewhat limited and introverted lives. The best-known ones were in the Trebizond region of north-eastern Turkey; in Crete; in Cyprus; and in parts of Albania and Kosovo (Kosova).[1] What the members of all these communities had in common was that they adhered outwardly to Islam – they made the profession of faith, they went to mosques, their menfolk were (in most cases) circumcised, they bore Muslim names, and so on – while, in the privacy of their own communities, they preserved the rudiments of a Christian faith, saying Christian prayers, having their children baptised if possible, observing Christian feasts and fasts, and performing other Christian rituals. The term 'communities' is emphasised here, because crypto-Christianity is very much a phenomenon of social religious life: individuals may have hidden their faith at all sorts of times and places, but the sort of crypto-Christianity I am concerned with here is a form of life, a tradition, something that can only be sustained and transmitted by a community.

The crypto-Christians of Kosovo form the most distinctive group out of all the Ottoman crypto-Christians, because their form of Christianity was Roman Catholic. But of course the first thing that strikes one about the whole phenomenon of crypto-Christianity is that so many of these communities sprang up, probably without any knowledge of the existence of other such communities in other parts of the Empire, and underwent quite similar patterns of development. Before considering the distinctive history of the Kosovo crypto-Christians, therefore, it is worth looking at some of the general conditions of religious life in the Ottoman-ruled Christian or former Christian territories. For much of this century – and certainly since the publication of F. W. Hasluck's marvellous *Christianity and Islam under the Sultans* in 1929 – studies of religious life in the Ottoman Empire, especially in the Ottoman Balkans, have paid great attention to the ways in which Islam and Christianity interacted and became elaborately intertwined. A much fuller and richer picture has emerged, which makes the Manichaean vision of the old religious-nationalist historiography (oppressed national churches versus oppressive Islam) seem crude and quite outmoded. Nevertheless, there is a danger that some key distinctions may get lost or blurred in this new, complex, mixed-up, parti-coloured world of Ottoman religion, in which everything seems to have been merging into everything else in a sort of warm syncretist soup. The phenomenon of crypto-Christianity may also get rather blurred when looked at in this way, becoming mixed up with other types of what might in general be called religious amphibianism; so I should like to begin by making some distinctions, and isolating a number of factors or conditions which I think are essentially different from crypto-Christianity, even though they may have helped to provide an environment of amphibianism in which crypto-Christianity could survive or flourish. The examples I shall give here are drawn mainly from the religious history of Kosovo, but they apply much more generally to conditions throughout the Ottoman Balkans. I want to distinguish three factors: the first is social coexistence; the second is religious syncretism; and the third is theological equivalentism.

The social coexistence of Christians and Muslims that matters here is not the general co-presence of the two faiths in towns or country areas, but the close coexistence of people living in such social intimacy that they could not avoid experiencing, and even in some ways sharing, the ritual acts or religious observances of each other's faith. The strongest form this took was the coexistence of religions within a single family. This was something that happened, typically, in either of two ways: a member of a Christian family converted to Islam (usually, a young man, who could then become the head of a Muslim sub-branch of the family); or there was intermarriage. Many examples of this sort of intimate coexistence could be given from all over the Balkans; let me give a few from the Kosovo region.

The modern Franciscan mission to northern Albania started in the 1630s, and in 1637 one of its members, Fra Cherubino, went into

Kosovo. At one house in a village outside Gjakova (Djakovica) he and his companion were welcomed with the words: 'Come in, Fathers: in our house we have Catholicism, Islam and Orthodoxy'. In shocked tones Fra Cherubino reported to his superiors: 'They seemed to glory in this diversity of religions, as if they were wiser than the other people of this world.'[2] In the clan system of the northern Albanians, loyalties to the family, the vëllazëri (bratstvo, group of families) or the clan were always stronger than the claims of religion: some clans, such as the Krasniqi and the Berisha, divided into Catholic and Muslim branches, without any diminution of fellow feeling.[3] In the Thaçi clan of north-eastern Albania there were three branches: one fully Catholic, one fully Muslim, and one Catholic but non-pork-eating. The clan tradition was that there had been three brothers in the late seventeenth or early eighteenth century: the eldest had converted to Islam, the second had remained Catholic, and the third, while remaining Catholic, had stopped eating pork out of deference to the eldest.[4] Whether this oral tradition was anything more than a fanciful rationalisation can only be guessed at; but the example does suggest that close social coexistence could lead to the blurring of some religious distinctions. In the 1930s and 1940s Mirko Barjaktarović studied a number of Albanian *zadrugas* (extended family farms) in Kosovo which contained mixtures of Catholics and Muslims; not surprisingly, he found a degree of syncretism in their religious practices. The Muslim members of the family would assist the Catholics in cutting the badnjak (Yule-log), and would also attend prayers on saints' days; and the Catholics would take part in the Bajram celebrations. When they had feast-days together the Catholics would not eat pork or drink wine, but all of them, Catholic and Muslim, would drink *raki* (brandy).[5]

Obviously, the coexistence, however close, of two distinct religious identities is not the same as crypto-Christianity. Each person in one of these mixed families was either one thing or the other. The only direct way in which this form of life may sometimes have created a small-scale simulacrum of crypto-Christianity was in the opportunity it may have given to some of the Christian members of the family to pass themselves off as Muslims when they travelled elsewhere, given that they had an unusually intimate knowledge of Muslim customs and religious observances. It was also quite common in these mixed *zadrugas* for the Catholic members to be given, in addition, Muslim names, which they used within the family and could use in the outside world if it were to their advantage to do so.[6] More generally, the use of Muslim names by Catholic Albanians in the Malësi (highlands of northern Albania) was in fact widespread until well into this century; it may originally have been just a form of camouflage for their dealings with the outside world, but it became a cherished tradition, and continued long after the need for such protection had passed.[7] Of course the principle of camouflage is the essential principle of crypto-Christianity; but crypto-Christianity involves a much more far-reaching form of mimicry than the mere use of a name for limited social purposes.

Religious syncretism provides my second form of amphibianism. The general phenomenon here is very well-known: at the level of folk-religion, many practices were shared between Muslims and Christians, for a range of purposes including the divinatory, medical and apotropaic. Some forms of folk-religion did not require the assistance of priests or imams: examples would include the private celebration of saints' days (Muslims in northern Albania would light a candle on St Nicholas's day), the celebration in Muslim homes of a 'slava' or family patron saint's day, and the ritual of the *badnjak* at Christmas-time. And some of these observances would be carried out by Muslims and Christians together: mixed Muslim–Christian villages in northern Albania, for example, would have joint celebrations of St Nicholas's day.[8] (There is, incidentally, a tradition of argument by Serbian ethnographers which claims that all these practices demonstrate that the Albanians of Kosovo and northern Albania were originally Serbs, as if only people who had started with pure Serbian blood in their veins could do any of these things. Such an argument misunderstands the whole nature of syncretism; it also misrepresents the origins of many of these practices, all of which can be found in a much wider catchment area and some of which have obviously pre-Christian and pre-Slav origins.)

Other forms of syncretism involved recourse to the priests and clergy, or the shrines and religious buildings, of the other faith. Thus Christians might go to the *türbe* (tomb) of a holy dervish sheikh, or ask a dervish to read the Koran over them when they were sick; and Muslims might visit churches, and ask priests for blessings, holy water or the administration of sacraments. Nineteenth-century travellers noted Albanian Muslims going to visit the monastery of Dečani; a Bosnian Franciscan who travelled through northern Albania in 1907 observed in one village that 'On Sunday, five or six Muslims came to church, and a bula knelt below the altar during Mass.'[9] For the more or less magical purposes of folk-religion, holy oil (chrism) and holy water were especially prized. Christian baptism was held in high esteem by many Muslims, for a variety of magical purposes: they thought it would give them a longer life, that it would guard them against mental illness, that it would protect them from being eaten by wolves, and – a strange idea, but a very widespread one – that it would prevent them from smelling like dogs.[10]

The reports sent back to Rome by the senior Catholic clergy in Kosovo often show signs of irritation at this magical use of Christian practices. The eighteenth-century Archbishop of Skopje, Matija Mazarek, reported wearily on one occasion that he had been summoned by the governor of Novo Brdo (Novobërda), whose wife was suffering from a certain curious female malady, and that he had had to bless oil and water for her; more generally, he complained that in order to placate the local Muslims he had to 'visit their sick, and even exorcise and bless their animals'.[11]

And yet these misgivings about folk-religion did not stop the Catholic Archbishops of Skopje (i.e. in effect, the archdiocese of Kosovo) from presiding over one of the most dramatic examples of syncretism in the whole of the Ottoman Balkans. This was the great two-day festival on the summit of Mt Paštrik (Pashtrik), a few hours walk to the west of Prizren, to celebrate the Assumption of the Virgin Mary. The mountain had three peaks: more than 1000 people would gather there, and say vespers on the two lower ones. Then, in the words of the visitation report of Archbishop Pjetër Bogdani in 1681: 'they spend all night there, with drums, whistles, dancing and singing. After midnight they begin a mixed procession – Muslims, Serbians and Greeks with lighted wax candles, their length proportionate to each person's age. They walk round the peak of the highest mountain for three hours in bare feet (with some of the leading Muslims on horseback).' In the morning, the Archbishop held a service on the mountain-top, preaching in Albanian to 'a numberless crowd of all sorts of people'; he was taken to lunch afterwards by the Orthodox Bishop of Prizren.[12] This cult of a mountain-top was evidently a pagan survival; what makes it a peculiarly fascinating example of syncretism is the fact that in later periods it was subsumed under two different interpretations, Muslim and Orthodox: one of the many tombs of the legendary Muslim holy man Sarí Saltík was located there, with celebrations on the day of Ali, the son-in-law of the Prophet, 2 August; and the Orthodox in Prizren in the late nineteenth century had their own version of this story, in which the grave of St Pantaleimon was on the summit – also with appropriate celebrations taking place over a whole night in the summer.[13]

Syncretism, once again, is clearly a different phenomenon from crypto-Christianity. A Christian who visited the türbe of a Muslim holy man to cure an illness was not pretending to be a Muslim; more importantly, a Muslim who had his children baptised may not have had any idea in his head of creating a secret Christian identity for them. All we can say is that in indirect ways the common or borrowed practices of syncretism may have helped to sustain an environment in which it was easier for crypto-Christianity to exist – in such an environment, crypto-Christians did not always need to engage in elaborate subterfuges in order to enter churches, have contacts with priests, or engage in some at least of the festivals and other practices of their Christian faith.

The third form of amphibianism I wish to look at very briefly is what I call 'theological equivalentism'. The term 'theological' is used here deliberately, as opposed to 'religious'. Most modern studies of religious life in the Ottoman Balkans, and particularly of its Islamicisation, deal almost exclusively with religion as a social phenomenon, and attribute conversion purely to social or economic factors. I should like to put in a small plea for the role of theological arguments, which have been almost pushed out of the picture. Of course most peasants had only the haziest ideas of Christian – or Islamic – doctrine; of course social and economic factors were important, and probably predominant in most cases. But when reading accounts by seventeenth-century Catholic clerics of their activities in Albania and Kosovo, one frequently finds that they appealed to theological arguments either to stop people from converting to Islam, or to reconvert them if they had already done so; it does not seem unreasonable to suppose that hodjas and dervishes might also have used theological arguments for their own purposes.

The argument which I describe as 'theological equivalentism' stated that both Islam and Christianity were equally valid ways to salvation. I use the clumsy word 'equivalentism' rather than more familiar terms such as 'ecumenism' or 'latitudinarianism', because those terms have too specifically Christian or intra-Christian meanings; another term used in seventeenth-century Christian discourse, 'indifferentism', also has too narrow a focus, as well as suggesting overtones of coldness or even disbelief, which need not apply in this case. It was Muslim proselytisers, not cold or irreligious people, who had recourse to the tactic of theological equivalentism, arguing that Christianity and Islam were hardly opposed to each other at all. In the words of one Francisan report from the mid-seventeenth century:

> Those impious people also said that the difference between them and the Christians was small; 'After all', they said, 'we all have only one God, we venerate your Christ as a prophet and holy man, we celebrate many of the festivals of your saints with you, and you celebrate Friday, our festive day; Mohammed and Christ are brothers...' And

this error was so widespread, that in the same family one person would be Catholic, one Muslim and one Orthodox.[14]

Similarly, a report of 1650 noted that 'the Muslims preach to them that everyone can achieve salvation in his own religion'.[15] This argument, which seems at first sight to strengthen the case against changing one's faith, was in fact a subtle first step towards converting Christians to Islam: the important move was to get them to accept the idea that Muslims too would achieve eternal life. And on a slightly different tack, Archbishop Mazarek reported, in 1760, on what he called the 'fine arts and stratagems' of the Muslims in trying to persuade Catholics to convert: they argued, he said, that the Gospels did in fact teach the doctrines of Islam.[16] This may be theologically a different position from strict equivalentism; but it has the same effect of downplaying the importance, or even the possibility, of doctrinal conflict between two religions.

This attitude is clearly conducive to a kind of religious amphibianism; it can be combined – as the quotation from the Franciscan report illustrated – with both syncretism and intimate social coexistence. But once again, although this may have contributed to forming an environment in which it was easier for crypto-Christianity to exist, there is an essential difference in principle between this sort of easy-going religious amphibianism and crypto-Christianity itself. Crypto-Christianity must surely have been based on the idea that the two faiths were not really equivalent at all, but radically opposed, and that one of them (the secretly held Christian faith) gave salvation, while the other did not.

Having made these distinctions, let us now turn to the history of crypto-Christianity in Kosovo. Some previous writers on this subject have offered as evidence of crypto-Christianity things for which – if my distinctions are correctly drawn – there may be quite separate explanations: the baptism of the children of Muslim parents, the celebration of Christian festivals in Muslim homes, the use of Muslim names by Christians, and so on. Nevertheless, there is quite strong evidence of genuine crypto-Christianity, which emerges in the seventeenth century and continues into the twentieth.

The origins of this phenomenon may go back earlier than that, but the evidence simply does not allow us to say with certainty that crypto-Christianity was present before the seventeenth century. The earliest potential evidence comes from the Ottoman report of 1568, which noted disapprovingly that Muslim villagers in the Debar (Dibra) area were taking their new-born children first to the priest, who gave them Christian names, and only afterwards to the Muslim clergy.[17] Without further details, we cannot tell whether this was proper crypto-Christianity or merely a quasi-magical syncretist practice. Some of the earliest visitation reports from northern Albania do allude to what might be called a crypto-Christian attitude: a report of 1603 mentions people who think they can profess Islam and at the same time 'retain the Christian faith in their hearts'.[18] People who had converted to Islam for mainly prudential reasons might well have thought that they remained Christians at heart. This is a phenomenon of first-generation converts, an attitude which may either have survived, or have died with them; only when the attitude has become transmitted down the family, and established as a tradition, can we speak confidently of the practice of crypto-Christianity.

The Archbishop of Bar (Tivar, Antivari), Marin Bizzi, described in a visitation report of 1618 what does seem like an initial phase of crypto-Christianity: some of the Catholics of northern Albania, he said, 'profess outwardly the Muslim religion...while retaining the Christian faith only in their hearts; and on that assumption, they demanded in vain from the Archbishop on his visitations that he should issue a decree, telling the parish priests to administer secretly to them the sacraments of penitence and the eucharist'.[19] But if this was an early sign of crypto-Christianity, then it cannot have been a common or an established phenomenon at this stage: when Bizzi issued a detailed set of instructions to his priests, including, for example, special requirements to stop the misuse of the sacraments for folk-religious purposes, he made no mention of crypto-Catholicism.[20]

Another report, made by the parish priest of Prizren, Gregor Mazrreku, in 1651, described a very similar request:

> Some of the men (and there are very many of these) say: 'We are Christians in our hearts, we have only changed our religious affiliation to get out of paying taxes which the Muslims imposed on us' and the reason they say...'dear Reverend, come and give us confession and Holy Communion secretly'. But I have not done this up till now, nor does it seem right to me...[21]

Clearly the phenomenon was growing by now. The first use by a Catholic priest in this area of a phrase meaning 'crypto-Christians', 'christiani occulti' or hidden Christians (in fact he meant 'hidden Catholics' – only Catholics were called 'Christians' by the Catholic clergy, while the

Orthodox were called 'Schismatics' comes in a report of 1672 by Shtje-fën Gaspari, describing the Pulat region of northern Albania. He says that the men of this area publicly converted to Islam twenty years ago:

> They are called by Muslim names in the presence of the Muslims, they eat meat and foods forbidden by our holy faith (i.e. during Catholic fasts), but when they are not observed by the Muslims they go to church to hear the Mass, to confess and to receive communion from a certain priest, Martin Politi, who goes sometimes to that region, with the bishop's permission, in order that the holy sacraments may be administered to these people, whom he calls hidden Christians.[22]

Just two years after that report was written, the first formal decree against crypto-Catholicism was issued, by Andrija Zmajević, the Archbishop of Bar. He ordained that Holy Communion should no longer be given to any Catholics who made a public profession of 'infidelity' and sought to hold their Christian faith 'occulte', in a hidden way, and damned with anathemas any priest who disobeyed this order.[23] Clearly, crypto-Catholicism was by now an established phenomenon.

Going back to those two reports, by Gaspari in 1672 and Mazrreku in 1651, we can get some further clues about how and why it was established. Note, first of all, that both reports specifically say that it was the men who had converted to Islam. Men-only conversion was a common expedient, reflecting the fact that the main motive for conversion in these cases was to escape the additional taxes on Christians, which were levied only on the male members of the family. The main tax was the *cizye* or haraç, which was paid, under normal circumstances, only by non-Muslims; some other taxes had differential rates, such as the ispence or resm-i çift; and there is evidence that at some times in the seventeenth century, when the Ottoman Empire was at war with Catholic powers, special extra taxes or charges were levied on the Catholic community. (There were probably other pressures specifically against Catholics, especially after the two abortive attempts by Catholic bishops to assist the conquest of Shkodra (Scutari) by Venetian forces in the 1640s.[24] These, like the later Austrian invasion of 1689, provoked a fiercely anti-Catholic policy on the part of the Ottoman authorities in the northern Albanian region: this may help to explain why it was that crypto-Christianity developed among the Catholics of Kosovo, but not, so far as we know, among either the Orthodox of Kosovo or the Catholics of other regions such as Bosnia.) Visitation reports frequently describe cases of whole Catholic communities engaging in men-only conversion. In 1637, for example, Gjergj Bardhi noted that the inhabitants of a village near Gjakova had divided in this way; in 1651 Gregor Mazrreku wrote that all the men in Suha Reka (Suva Reka), where there had previously been 160 Catholic households, had gone over to Islam, but that thirty-six or thirty-seven of their wives remained Catholic; and twenty-one years later Shtjefën Gaspari found 300 Christian women but no Christian men at all in the Has district west of Gjakova.[25] Gregor Mazrreku reported that the men who had converted to Islam 'do not want to take Muslim women as their wives, but Christian ones instead, saying, "it's so that in this way the name of Christian will not die out completely in my house." '[26]

For Muslims to take Christian wives was quite a common practice in the Ottoman Empire, so it would not have aroused any suspicions on the part of the authorities or the Muslim clergy. Under Islamic law it is permissible for a Muslim man to marry a Christian or Jewish woman, though not for a Muslim woman to marry a non-Muslim. An anonymous Venetian account of the Ottoman Empire in 1579 described the practice as follows: 'The Turks (i.e. Muslims) take Christian wives without demur, since their law permits it. The male child, at the father's request, is made a Muslim, and the female ones, at the mother's request, become Christians; however, the girls will be brought up as Muslims if the father desires it.'[27] Apparently, therefore, this system of religious differentiation by sex could be continued down the generations. It may thus have been an important mechanism in making crypto-Catholicism possible. With women in the family who were officially Christian, it became possible for priests to enter their homes in order to minister in secret to the menfolk as well. These home visits were an important part of the job of a parish priest: as one priest reported in a deposition to the Vatican in 1728, 'there are 237 Catholic households with 4,695 Catholic souls scattered among the houses of the Muslims who are given the sacraments by us.'[28]

If sexual differentiation was one key mechanism in the development of crypto-Catholicism, then the other most important mechanism was priestly complicity. It is no coincidence that Gaspari's report of 1672, the first definite description of fully functioning crypto-Christianity, was also the first to confirm that these 'hidden Christians' were being given the sacraments: quite simply, without the co-operation of the priests, crypto-Christianity could not function properly at all. The prohibition issued by Andrija Zmajević in 1674 seems to have had little effect; the administration of the sacraments continued. In terms of official doctrine, Zmajević was of course quite correct: making a public denial of Christ, such as implied by conversion to Islam, was a mortal sin, and the sacraments could not be administered to anyone who did not repent of it and return to the Church. A specific decree against crypto-Christianity had been issued by the Congregatio de Propaganda Fide (in response to an enquiry from a missionary in North Africa) as early as 1630.[29] It was precisely in that period, the early seventeenth century, that the Catholic Church developed a sophisticated theory of 'mental reservations' to protect Catholic priests who were questioned about their religious identity in Protestant countries; but although the theory covered various sorts of misleading answers to questions, it could not stretch far enough to include an explicit profession of faith in another religion.[30]

Twenty-nine years after Andrija Zmajević issued his decree, his nephew and successor-but-one as Archbishop, Vicko Zmajević, held a provincial synod (in 1703) which repeated and in some ways strengthened the provisions against crypto-Catholicism. Not only did the decrees of this synod say that those who professed Islam must be denied sacraments; they also said that even without making an official profession of faith in Islam, if someone lived 'Turcico more', in the Muslim way, using a Muslim name, and eating on Catholic fast-days, then that person too should be denied the sacraments. Other decisions of this synod were that a Christian girl who married a Muslim must be denied the sacraments (though she could continue to receive them if the man had been a Christian when she married him), and that although it was permitted to avoid declaring one's Christian faith when asked by a private person, if asked by public authority one was obliged to profess, even at the risk of death.[31] Further prohibitions, along similar lines, were issued by the Holy Office (the Inquisition) in 1724 and 1730.[32]

Despite all these dire prohibitions, however, the crypto-Catholics continued to receive the sacraments from their local priests. That they did so is hardly surprising: the two Austrian invasions of Kosovo, in 1689 and 1737, unleashed waves of Ottoman anti-Catholic hostility in the region, and must have prompted many prudential conversions to Islam. The fullest description of how crypto-Catholicism worked comes from just after that second Austrian fiasco: it derives from a report to Rome by the Archbishop of Skopje, Gjon Nikollë (Ivan Nikolović) in 1743. (Unfortunately the section of his report describing the crypto-Catholics has not yet survived; but a long summary of it has.) Nikollë clearly stated that the bishops and priests did administer all the sacraments to these people, in open violation of the decrees of the provincial synod.[33]

He wrote that the Catholics in the cities, and in some villages, professed their faith openly, paid the tribute and suffered persecution: the others, scattered in the villages, 'maintain the Christian faith in a hidden way' ('occultamente'). These, he said, 'profess the Christian faith in an internal way, but so secretly, that sometimes the father does not reveal himself as a Christian to his sons, or the sons to the father, and on their deathbeds they behave as and give themselves out to be Muslims'. He observed that some went to the mosques; in some cases they had themselves circumcised, in order not to be recognised as Christians, and they were buried as Muslims; but on the other hand they professed Christianity at home, baptised their sons, confessed, communicated, observed vigils and Lent, and said masses for the dead. Interestingly, he noted that 'their wives are for the most part publicly Christian, unless they are themselves the daughters of Muslims converted subsequently in the homes of their Christian husbands.'[34] Nikollë's remark about sons not revealing their Christianity to their fathers is puzzling; either it indicates some covert induction of sons into Christianity by Christian mothers, or it may perhaps be attributable to a pattern of rhetorical phrase-making taking over from the sense of Nikollë's argument. But the reference to fathers not telling their sons is interesting, in view of the comparison it prompts with what we know of crypto-Judaism in the seventeenth century: in the case of the 'Marranos' in Spain and Portugal, apparently, it was common for fathers to wait until their sons were aged twenty before beginning their induction into Judaism.[35]

When Archbishop Nikollë sent his report to Rome in 1743, he asked for further guidance on this whole problematic issue. All he got in response, from Pope Benedict XIV, was even more severe restatement of the official line in an encyclical of 1744. (Another encyclical, ten years later, also tightened up the official position on the use of Muslim names.)[36] The next-but-one Archbishop, Matija Mazarek, who held the post for most of his adult life (from 1758 to 1807), did make at least a partial attempt to enforce these decrees. In a report of 1760 he wrote that the crypto-Catholics used to receive the sacraments until Benedict XIV's encyclical; now they were rapidly turning to Islam. The saddest cases among them, he wrote, were those of women who had been converted secretly to Christianity after marrying crypto-Christian men, but who were now also refused the sacraments.[37] He also told the story of an eighteen-year-old Albanian who had come to a christening service conducted by Mazarek in Peć (Peja), had said he was a crypto-Christian and asked to be baptised. Mazarek told him to profess his faith openly, and he refused; so Mazarek denied him baptism. But Mazarek wrote that his own conscience was troubled, and that this policy was not only losing souls, but creating enemies: 'because they have been abandoned by us missionaries, they completely embrace Islam, and these people bear an incredible hatred, aversion and contempt towards us, and we suffer worse persecution from them than from the true and original Muslims.'[38] Relenting somewhat, when he visited a mainly crypto-Catholic village near Prizren during the same visitation, he did allow them the sacraments, on condition that they promised him four things: not to eat forbidden foods, not to circumcise, not to enter mosques and not to allow Muslim burial for their dead.[39] Some such compromises, it seems, must have continued to be applied in practice; otherwise it is very hard to see how the phenomenon of crypto-Catholicism could have continued, as it did, for several more generations. A report by Archbishop Bogdanović in 1846, for example, gave quite detailed statistics, not only for Catholic families in the archdiocese, but also for crypto-Catholic ones: 128 of them in the parish of Prizren, for example, 150 in the Karadak (Skopska Crna Gora), 57 in the parish of Janjevo, 165 in the parish of Peć, and so on, up to a total of 500 families.[40] Twenty years later a French consular official in Salonica was able to report that the archdiocese of Skopje contained precisely 5847 open Catholics, and 4735 crypto-Catholics, the latter in 665 families.[41] Such statistics must reflect continual contacts between the crypto-Catholics and the parish clergy.

It was in the 1840s that the first attempts were made by crypto-Catholics to make a public profession of faith on a legal basis. The great reform decree of 1839, the Hatt-i Şerif of Gülhane, had included general promises of equal rights for all subjects, regardless of religion; five years later, after much pressure from the British Ambassador in Istanbul, a declaration was made that Muslim converts from Christianity who wanted to revert to the Christian faith would no longer be subject to the death penalty.[42] In the following two years attempts were made to introduce the new conscription system, and new taxation, to Kosovo: both forms of innovation were resisted, and a large military force was sent to crush the rebellion. When the Ottoman army commander imposed order in Gjakova in the summer of 1845 and began conscripting the local men, the call-up was applied only to Muslims. (This was the traditional Ottoman policy, although the new army law of 1843 did in principle make Christians liable for military service too.)[43] A group of crypto-Catholics from nearby villages went to the Catholic church in Gjakova and declared themselves as Christians; and the same happened soon afterwards in Peć. Seventy of these men then made a public declaration of Christianity in the army camp. This was an act of some bravery: everyone in this part of Kosovo would have remembered the fate of a group of Catholics in Rugova (Rugovo), a village west of Peć, who had Muslim names but professed Catholicism openly, and were executed by the local pasha in 1817.[44]

The crypto-Catholics who declared themselves in 1845 were also put into prison, but the army commander did at least order an inquiry, at which the Catholic priests were allowed to make their case. When they argued that these people had only simulated Muslim beliefs in order to avoid oppression, the local Ottoman officials raised the counter-example of the Fandi, a warlike Catholic clan (and sub-branch of the Mirdita) who had moved into the area west of Gjakova in recent decades. 'Why is it that the Fandi have never been disturbed in their religious affairs?' they asked, rhetorically. The parish priest of Gjakova replied that they were a powerful clan who could defend themselves, whereas his crypto-Catholics were 'scattered among completely Muslim villages'. After six weeks in gaol (during which two died of dysentery), the Catholics were released, on payment of a fine. Fifty more heads of family promptly came out as Christians, bringing the total to more than 150.[45]

Later that year, and in the spring of 1846, more crypto-Catholics declared themselves in the villages of the Karadak. The treatment they received was harsher: both they and their priest were thrown into prison, and when they failed to pay the allotted fine (despite, or perhaps because of, the confiscation of all their property), they were sent into exile in Anatolia. Roughly 150 people (twenty-five families) underwent this punishment, with their priest; twenty of them perished on the way to Anatolia, and at least seventy more died in exile before they were finally permitted to come home two years later.[46] Similar problems involving the conscription of crypto-Catholics arose again in 1849 and 1850.[47] But the position of the Catholic Church did generally improve, and the leverage exerted by Western diplomats in Istanbul was enormously increased during the period of the Crimean War, when the Ottoman state was allied with European powers against Russia. In 1856 another important reform decree was issued declaring full equality of rights among Muslim and non-Muslim subjects and full freedom of religion; this decree, which included a remarkable clause against hate-speech (forbidding officials to use words or expressions 'tending to make one class of my subjects inferior to another class on account of religion, language or race'), also confirmed that apostasy from Islam was no longer punishable by death.[48] After this, the general persecution of the Kosovo crypto-Catholics seems to have ceased. Some declared themselves publicly as Catholics soon after 1856; but the Catholic archbishop noted that many remained in their crypto-Catholic state in his report of 1872.[49]

The most persistent community was the small group of crypto-Christian villages in the Karadak. There is some evidence to suggest that the most famous historical figure from this region, the military chief Idriz Seferi (who took part in every Albanian revolt from the League of Prizren in 1878 to the kaçak rebellion of the 1920s) was in fact a crypto-Christian.[50] During the period immediately after the First Balkan War, when Serbian and Montenegrin rule was forcefully imposed on Kosovo, there were reports of pressure being put on crypto-Catholics to convert to Orthodoxy; and in the 1920s it was reported that some of them declared themselves as Muslims in order to emigrate to Turkey.[51] Quite a number did come out as public Catholics in the 1920s and 1930s, but the ethnographer Atanasije Urošević wrote as late as 1935 that the crypto-Catholics were still very 'secretive'.[52] Recent research by Ger Duijzings suggests that it was the active policy of the Bishop of Skopje, Ivan Franjo Gnidovec (who was bishop between 1924 and 1939), to put pressure on the crypto-Christians to become full Catholics.[53] After all, the old decrees anathematising the adoption of Muslim practices were still valid, and however oppressive the policies of the Yugoslav state may have been during that period, they certainly did not involve pressure to conform to Islam. And yet it is reported that the traditions of crypto-Christianity had become so ingrained that the practice has continued, partly because of close ties with other Muslim families, until this day. Perhaps some element of genuine bi-confessionality has evolved. If so, the paradoxical conclusion must be that crypto-Christianity has been saved from extinction only because it has changed, during the final period of its history, into something else, ceasing to be merely a type of camouflage – a sincerely practised faith conducted in secret behind a simulated one – and becoming, after all, amphibious in the fullest sense: an equal and parallel commitment to two distinct forms of religious life.

Notes

1 See J. G. von Hahn, *Albanesische Studien*, 3 vols (Jena, 1854), vol. 1, p. 36; R. L. N. Mitchell, 'A Muslim-Christian Sect in Cyprus', *The Nineteenth Century and After*, 63 (Jan–June 1908), pp. 751–62; F. W. Hasluck, *Christianity and Islam under the Sultans*, ed. M. M. Hasluck, 2 vols (Oxford, 1929), vol. 2, pp. 469–74; R. M. Dawkins, 'The Crypto-Christians of Turkey', *Byzantium*, 8 (1933),

pp. 247–75; K. Amantos, *Scheseis Ellênôn kai tourkônapo tou endekatour aiônos merchri tou 1821* (Athens, 1955), pp. 193–6; R. Kiszling, 'Glaubenskämpfe in Albanien um die Jahrhundertswende', *Mitteilungen des österreichischen Staatsarchvis*, 12 (1959), pp. 426–32; S. Skendi, 'Crypto-Christianity in the Balkan Area under the Ottomans', *Slavic Review*, 26 (1967), pp. 227–46; A. Bryer, 'The Crypto-Christians of the Pontos and Consul William Palgrave of Trbizond', in his *Peoples and Settlements in Anatolia and the Caucasus, 800–1900* (London, 1988), item XVII; K. Photiades, *Hoi exislamismoi tês Mikras Asias kai hoi kryptochristianoi tou Pontou* (Salonica, 1993); G. Andreadis, *The Crypto-Christians: Klostoi: Those who Returned; Teneseur: Those who Have Changed* (Salonica, 1995).

2   D. Gubernatis and A. M. de Turre, *Orbis seraphicus; historia de tribus ordinibus a seraphico patriarcha S. Francisco institutis deque eorum progressibus per quator mundi partes [...] tomus secundus*, ed. M. a Civetia and T. Domenichelli (Quaracchi, 1886), p. 492b.

3   G. Valentini (ed.), *La legge delle montagne albanesi nelle relazioni della missione volante 1880–1932* (Florence, 1969), p. 129; M. Krasniqi, *Lugu i Baranit: monografi etno-gjeografike* (Prishtina, 1984), p. 44, with examples from Kosovo.

4   Valentini (ed.), *La legge delle montagne*, p. 205.

5   M. Barjaktarović, 'Dvovjerske šiptarske zadruge u Metohiji', Srpska Akademija Nauka, *Zbornik radova*, vol. 4 (Etnografski Institut, vol. 1) (1950), pp. 197–209; here pp. 204–6.

6   *Ibid.*, pp. 200–8.

7   F. Siebertz, *Albanien und die Albanesen: Landschafts- und Charakterbilder* (Vienna, 1910), p. 107; Valentini (ed), *La legge delle montagne*, p. 24; C. Libardi, *I primi moti patriottici albanesi nel 1910–1911–1912, specie nei Dukagnini*, 2 parts (Trento, 1935), part 1, p. 33; C. S. Coon, *The Mountains of Giants: a Racial and Cultural Study of the North Albanian Mountain Ghegs* (Cambridge, Mass., 1950, p. 36 (giving examples from the Nikaj and Merturi clans in 1929).

8   L. Mihačević, *Po Albaniji: dojmovi s puta* (Zagreb 1911), p. 10; see also A. Dumont, *Le Balkan et l'Adriatique: les bulgares et les albanais; l'administration en turquie; la vie des campagnes; le panslavisme et l'hellénisme*, 2nd edn (Paris, 1874), p. 289.

9   J. G von Hahn, *Reise von Belgrad nach Salonik* (Vienna, 1861), p. 48 (Sveti Prohor); G. Muir Mackenzie and A. P. Irby, *Travels in the Slavonic Provinces of Turkey-in-Europe*, 3rd edn, 2 vols (London, 1877), vol. 2, p. 88 (Dečani); Mihačević, *Po Albaniji*, p. 30. The 'bula' is a female Muslim religious teacher who also officiates at ceremonies for women.

10   Hasluck, *Christianity and Islam*, vol. 1, pp. 31–6; de Gubernatis and de Turre, *Orbis seraphicus*, p. 588b; G. Stadtmüller, 'Das albanische Nationalkonzil vom Jahre 1703', *Orientalia christiana periodica*, vol. 22 (1956), pp. 68–91, here p. 71; S. Vyronis, 'Religious Changes and Patterns in the Balkans, 14th–16th Centuries', in H. Birnhaum and S. Vyronis (eds), *Aspects of the Balkans: Continuity and Change* (The Hague, 1972), p. 151–76, here p. 174; Mitchell, 'A Muslim-Christian Sect', pp. 752.

11   Archivio della Sacra Congregazione de Propaganda fide, Rome [henceforth abbreviated: ASCPF], SOCG 895 (report of 17910, fo. 75^r: 'certa curiosa mala-

tia donesca', and SOCG 872 (report of 1785), fo. 130^r: 'visitare i loro malati, ed anche nell' escorcizare, e benedire i loro animali'.
12 ASCPF SOCG 482, fos. 288–9: 'ivi tutta la notte se la passano, con Tamburi, Siffare, balli, e Canti. Passata la mezza notte principiano una confusa Processione Turchi, Serviani, e Greci con candele di cera accese à misura della propria vita longhe. Circondano la Cima del più alto monte per spatio di tre hore à piedi scalzi alcuni de turchi principali à cavallo...'
13 N. Clayer, *L'Albanie, pays des derviches: les ordres mystiques musulmans en Albanie à l époque post-ottomane (1912–1967)* (Berlin, 1990), pp. 22, 171 (Sarí Saltík); B. Nušić, *S Kosova na sinje mora: beleške s puta kroz Arbanase 1894.godine* (Belgrade, 1902), p. 35 (Pantaleimon). On Sarí Saltík more generally see H. T. Norris, *Islam in the Balkans: Religion and Society between Europe and the Arab World* (London, 1993), pp. 146–57.
14 de Gubernatis and de Turre, *Orbis seraphicus*, p. 590. The Catholics did celebrate Fridays because of the Muslims, and their bishops had promoted the cult of 'Sancta Veneranda' in order to assimilate this practice (p. 591a).
15 I. Zamputi (ed.), *Relacione mbi gjendjen e Shqipërisë veriore e të mesme në shekullin XVII*, 2 vols (Tirana, 1963–5), vol. 2, p. 398. A seventeenth-century German author also noted that those who live among Muslims and observe their daily piety and good works 'will come to think that they are good people and will very probably be saved': T. W. Arnold, *The Preaching of Islam: a History of the Propagation of the Muslim Faith*, 3rd edn (London, 1935), p. 165–6.
16 ASCPF SOCG 792, fo. 153 ('fine arti, ed astuzie').
17 The document, in the state archives in Tirana, is discussed in P. Thëngjilli, *Renta feudale dhe evoluimi i saj në vise Shqiptare (shek. XVII–mesi i shek. XVIII)* (Tirana, 1990), p. 96, and N. Limanoski, *Islamizacijata i etničkite promeni vo Makedonija* (Skopje, 1993), pp. 137–8.
18 I. Zamputi (ed.), *Dokumente të shekujve XVI–XVII për historinë e Shqipërise*, vols 1–3 (Tirana, 1989–90), vol. 3, pp. 50–1.
19 M. Jačov (ed.), *Spisi tajnog vatikanskog arhiva XVI–XVIII veka* (Belgrade, 1983), p. 38.
20 D. Farlati, *Illyrici sacri*, 8 vols (Venice, 1751–1819), vol. 7, pp. 120–3.
21 Zamputi (ed.), *Relacione*, vol. 2, pp. 440–2.
22 S. Gaspari, 'Nji dorshkrim i vjetës 1671 mbi Shqypni', *Hylli i Dritës*, vol. 6 (1930), pp. 377–88, 492–8, 605–13, vol. 7 (1931), pp. 154–61, 223–7, 349–55, 434–47, 640–4, 699–703, vol. 8 (1932), pp. 48–50, 98–104, 208–10, 265–7, 310–14; here vol. 8, p. 100.
23 Farlati, *Illyrici sacri*, p. 136.
24 See Arnold, *Preaching of Islam*, pp. 188–9; de Gubernatis and de Turre, *Orbis seraphicus*, pp. 564–6
25 Zamputi (ed.), *Relacione*, vol. 2, pp. 97, 440; Gaspari, 'Nji dorshkrim', vol. 6, p. 386.
26 Zamputi (ed.), *Relacione*, vol. 2, p. 442.
27 E. Albèri (ed.), *Relazioni degli ambasciatori veneti al senato*, 15 vols (Florence, 1839–63), series 3, vol. 1, p. 454.
28 Archivio Segreto Vaticano, Vatican City, Processus consistoriales, vol. 114, fo. 617: 'Domicilia fidelium sunt 237 et animae fidelium 4695, et multae aliae

animae fideles sunt dispersae inter domicilia Turcharum et sunt a nobis administrata sacramenta'.

29 Farlati, *Illyrici sacri*, vol. 7, p. 146. I am not sure whether this is identical with the decree referred to by L. Rostagno and dated by her 1628: 'Note sulla simulazione di fede nell' Albania ottomana', in G. Calasso et al., *La bisaccia dello sheikh: omaggio ad Alessandro Bausani islamista nel sessantesimo compleanno* (Venice, 1981), pp. 153–63; here p. 156.

30 On the theories see P. Zagorin, *Ways of Lying: Dissimulation, Persecution and Conformity in Early Modern Europe* (Cambridge, Mass., 1990), pp. 153–220; J. P. Sommerville, 'The New Art of Lying: Equivocation, Mental Reservation, and Casuistry', in E. Leites (ed.), *Conscience and Casuistry in Early Modern Europe* (Cambridge, 1988), pp. 159–84.

31 Farlati, *Illyrici sacri*, pp. 147, 148, 158; on this synod see also Stadtmüller, 'Das albanische Nationalkonzil'.

32 Rostagno, 'Note', p. 157.

33 ASCPF SC Servia 1, fo. 318.

34 *Ibid.*, fos. 330 ('mantengono occultamente la Fede Christiana'), 317 ('professano interiormente le fede Christiana, ma tanto nascostamente, che talvolta il Padre non si palesa ai Figli, ne i Figli al Padre, e nell' estremo si mostrano, e fanno credere per Turchi... Le Mogli di essi sono per lo più publicamente Christiane, purchè non siano Figlie di Christiani occulti, o di Turchi e convertite poi in casa de loro Mariti Christiani').

35 Zagorin, *Ways of Lying*, p. 56.

36 Farlati, *Illyrici sacri*, pp. 172–6, 188–9.

37 ASCPF SOCG 792, fo. 147v.

38 *Ibid.*, 145v–146r: 'come per esser abbandonati da noi Missionari, assolutamente abbracciano il Mahometanismo, e questi tali ci portano un' incredibile odio, aversione, et esecrazione, e la più grande persecuzione da essi patiamo, che dalli veri, et antichi Turchi'.

39 *Ibid.*, fo. 149v.

40 ASCPF SC Servia 4, fos. 238v–240r

41 É. Wiet, 'Mémoire sur le pachalik de Prisrend', *Bulletin de la société de géographie*, ser. 5, vol. 12 (1866), pp. 273–89; here pp. 36–45.

42 R. H. Davison, *Reform in the Ottoman Empire 1856–1876* (Princeton, NJ, 1963), pp. 36–45.

43 H. Kaleshi and H.-J. Kornrumpf, 'Das Wilajet Prizren: Beitrag zur Geschichte der türkischen Staatsreform auf dem Balkan im 19. Jahrhundert', *Südostforschungen*, vol. 26 (1967), pp. 176–238; here p. 181.

44 ASCPF SOCG 922, fo. 333 (Krasniqi report, 1820). In Sept 1820 Rome began a 'processo informativo' on the Rugova martyrs (fos. 314–15).

45 All these details are from ASCPF SC Servia 4, fos. 172–3 (report by Fra Dionisio d'Afragola), and fos. 178–80 (report by Gaspër Krasniqi): 'perchè i Fantesi non sono stati mai disturbati nei loro affari di Religione?'; 'dispersi tra i Villaggi tutti Turchi' (fo. 172v).

46 ASCPF SC Servia 4, fos. 209–10 (letter from Archbishop Bogdanović), 237 (visitation report by Bogdanović giving the figure of 158 people); G. Gjini, *Ipeshkvia Shkup-Prizren nëpër shekuj* (Zagreb, 1992), pp. 190–3; G. Gjergj-Gashi, *Martirët shqiptarë gjatëviteve 1846–1848* (Zagreb, 1994), pp. 61–5, 82–92 (with 148 names and other statistics).

47 ASCPF SC Servia 4, fos. 350–2, 368–70 (Bogdanović reports, 1849, 1850).
48 Davison, *Reform*, pp. 55–6; text in A. Séopov ['Schopoff'], *Les Réformes et la protection des chrétiens en Turquie 1673–1904* (Paris, 1904), pp. 48–54 (quotation p. 51).
49 Von Hahn, *Reise von Belgrad*, p. 83 (noting 17 families who had declared themselves recently at Letnica); Archbishop Bucciarelli, report of 1872 (ASCPF SC Servia 5, fo. 589v).
50 The Serbian writer and vice-consul Milan Rakić stated that it was common knowledge: Rakić, *Konzulka pisma 1905–11*, ed. A. Mitrović (Belgrade, 1985), p. 239. Bejtullah Destani tells me that he has received confirmation of this claim from a member of Idriz Seferi's family.
51 L. Freundlich, *Albaniens Golgotha: Anklageakten gegen die Vernichter des Albanervolkes* (Vienna, 1913), p. 27.
52 A. Urošević, *Gornja Morava i Izmornik*, Srpski etnografski zbornik, vol. 51 (Naselja i poreklo stanovništva, vol. 28) (Belgrade, 1935), p. 112; on the coming out of crypto-Catholics in the 1920s and 1930s see Gjini, *Ipeshkvia*, p. 153.
53 G. Duijzings, 'The Martyrs of Stubla', unpublished typescript. I am very grateful to Ger Duijzings for letting me see this work.

Part II
Orthodoxy

7

The Chilandariou Icon 'Mother of God Tricheirousa': History, Cult and Tradition

Vladeta Janković

The Orthodox Christian monastic tradition, which has its roots in Egypt, achieved its flowering on Mount Athos, the most prominent of the three peninsulas stretching from the massif of the Chalkidiki south to the Aegean Sea. The monastic communities created there evolved with time into a monastic state which still has a certain degree of territorial autonomy guaranteed by international conventions, within the framework of the Greek state. Known as the Holy Mountain, it has proved, through the ages, to be the chief cornerstone and bastion of Orthodox monastic life.

The number of monasteries on Mount Athos fluctuated over time but has stabilised in the last few centuries: in addition to a considerable number of 'sketes' and 'kellia' type of hermitages, there are twenty monasteries. Seventeen of these are Greek, while three belong to Slav monastic communities – Russian, Bulgarian and Serbian. According to the hierarchy of the Athos monasteries, last established in the sixteenth century, the five held in the highest esteem are also entrusted with the administration of the Holy Mountain. The first of these is the Great Lavra, the second Vatopediou, then Iviron, the fourth Chilandariou, and lastly Dionissiou. Of these, the only one which is not Greek is Chilandariou. For the last 800 years, it has belonged to Serbs and served for their worship. The anniversary of its foundation was celebrated in April 1999.

Before Chilandariou was established as a Serbian monastery in the twelfth century, there had been a smaller Greek monastic settlement of the same name on the site. The deserted land and derelict buildings of this original Chilandariou had been under the supervision of the great

Greek monastery, Vatopediou. The Serbian ruler Stefan Nemanja, together with his youngest son Rastko Nemanjić, then rebuilt Chilandariou.

As a very young man, Rastko left his father's court for Mount Athos where he took monastic vows. He was received into the monastery of Vatopediou and became known as the monk Sava. Stefan Nemanja, his father, the founder of the Serbian medieval state and the progenitor of the dynasty, abdicated the throne a few years later in favour of his second son Stefan, known as the First Crowned, and followed Rastko to the Holy Mountain where, as an old man, he too took monastic vows, assuming the name of Simeon. Owing to the great influence and authority which both men enjoyed at the Byzantine court, they succeeded in obtaining a Charter with a Golden Seal from Emperor Alexius III Angelus (1195–1203), bestowing on the Serbian nation 'for all time' the land and estates of the abandoned Chilandariou, with permission to build a monastery which would be there 'to receive men of the Serbian people who wished to dedicate themselves to the monastic life, unfettered by any external authority'. The original of this document, dated 1198, is preserved in the archives of Chilandariou.

On the strength of generous financial help dispensed by the new Serbian ruler Stefan the First Crowned, Simeon and Sava (Nemanja and Rastko) were able, in a very short time, to build a new monastery and provide it with all the indispensable necessities. Soon afterwards, on 26 February 1200, Nemanja died and was buried in the Chilandariou church, only to be disinterred several years later and his relics returned to Serbia where they were finally laid to rest in the monastery of Studenica. Sava himself left Mount Athos after a time, in order to become the first archbishop of the newly independent Serbian Church. As a great reformer and the spiritual father of the Serbian people, he was equally active as a diplomat, law-maker, writer and educator, until his death in 1235.

Both men, father and son, were later canonised. Saint Sava is still today the greatest Serbian saint and a powerful symbol of Serbian national identity, statehood and culture. Equally, in Chilandariou, his memory is still deeply revered, not only because he built the monastery and was the greatest benefactor of his own people, but also because he brought to the monastery its most sacred object, the icon of the Mother of God Tricheirousa or 'She Who Has Three Hands', whose cult, tradition and history are the subject of this chapter.

The Virgin of the Three Hands (Fig. 7.1) is indisputably held to be the holiest icon in all of Serbian Orthodoxy. Among the people there is a

Figure 7.1 The Virgin of the Three Hands.

profound faith in its miraculous powers and a wealth of beliefs and legends connected with it are fostered and cultivated in monastic circles. Consequently, it is no simple matter to pin down the facts surrounding its provenance and historical destiny. Here, as in many other cases, the church's account and the folk legends all frustrate and hinder a scholarly approach to the truth.

The following facts are reliable, however. The Virgin of the Three Hands is a processional icon, measuring 111 cm by 91 cm and on its reverse side there is a painting of Saint Nicholas. It belongs to the type known as 'Odigitria' (or Protectress of Wayfarers) and experts are generally agreed that it dates from the fourteenth century. It owes its strange name to the fact that below the Virgin's right hand, supporting the Christ-child, there is another hand, evidently added subsequently, and wrought in silver. The icon is banded with gold and inlaid with semi-precious stones. It has pride of place, standing on the spot traditionally reserved for the abbot – in front of the weight-bearing pillar on the right-hand side of the nave facing the altar.

The most widespread belief is that the icon was painted by the evangelist Luke himself and that in the eighth century it was in the possession of Saint John of Damascus. St John was a staunch defender of icons and their veneration at the time of the iconoclastic movement, whose chief instigator was the Byzantine Emperor Leo III Isaurian. Legend has it that Emperor Leo falsely accused St John of plotting an uprising against the Syrian Caliph Valid, who then punished the Christian defender of icons by ordering that his right hand be severed. According to legend, St John spent a whole night in prayer before the Icon of the Virgin, imploring her to heal his hand so that he might continue to write in the cause of holy icons and their protection. Exhausted with the pain and effort, he dropped off to sleep for a while and when he awoke, he saw that his hand was healed; only a thin red line remained visible, as a reminder of his suffering and a sign of the miracle which had taken place. As a mark of gratitude, St John then ordered that a hand be fashioned in silver and attached to the lower left part of the icon.[1] (This is, in fact, only a reflection of the very ancient custom of offering a votive gift to the deity, a representation of the part of the body which had been healed.) Ever since then, legend says, the icon has been known as She Who Has Three Hands.

According to hagiography, the story goes on to relate how St John then became a monk and remained in the monastery of Saint Sabbas of Jerusalem to the end of his life, inseparable from his miraculous icon, his protector and guardian. In addition to being the resting-place of St Sabbas, its founder (in the sixth century), the monastery was also the repository of his abbot's staff, which was fixed beside his grave. St John Damascene had prophesied that one day the son of a Tsar, bearing the name of the Saint (that is to say, Sabbas or Sava), would come from a distant land, whereupon the staff would fall to the ground. This man would be the chosen one to whom the fraternity was to entrust the staff and also the icon of the Three-Handed One, bequeathed to him by St John.

Five centuries later, in 1217, according to the same source,[2] the Serbian St Sava, son of a Tsar (though at that time still a pilgrim monk), arrived in Palestine. As he was paying his respects to the grave of the monastery's founder, the staff fell. The fraternity was sceptical, but as the same occurred again the following day, they conceded that the prophecy had indeed been fulfilled and handed over the abbot's staff and with it the icon of the Three-Handed Virgin. This is how St Sava the Serb came to bring the two holy relics to Mount Athos. The abbot's staff is kept today in the Cell of Transfiguration in Karyes, the administrative centre of the Holy Mountain, while the icon acquired its rightful place in the holy temple of the monastery of Chilandariou.

In addition to this partly sacred, partly folk tradition, there exists another purely ecclesiastical one, which has its source in monastic circles and refers to subsequent occurrences regarding the icon. The Three-Handed One remained in Chilandariou until the middle of the fourteenth century when the monastery was visited by Stefan Dušan, of the Nemanjić dynasty, who had declared himself Tsar of the Serbs and Greeks, and effectively ruled over the greater part of the Byzantine Empire at that time. Tsar Dušan, it is believed, removed the icon of the Three-Handed One, with or without the consent of the fraternity, and carried it off to Serbia. It is said that the icon was then placed in Studenica, 'The Mother of Serbian Monasteries', where it remained until the Turkish invasion in the fifteenth century. In order to hide and save their sacred treasures from the invading hordes, the monks of Studenica are said to have attached the icon to the back of a donkey and sent it off, 'to go where the will of the Virgin takes it'. In this way, it is alleged, the donkey alone covered the vast distance through all of Serbia and Macedonia to reach Mount Athos where, within a hundred yards of the gate of Chilandariou, it dropped dead from exhaustion. The monks immediately returned the icon to its accustomed place in the church and, on the spot where the donkey ended his journey, built a chapel. To this day, every year on 12 July, the icon is carried in procession out of the church to the chapel, where a service is held to commemorate the miraculous event.[3]

There is another curious tale associated with the icon of the Three-Handed Virgin. It tells of a dispute, which arose between the monks after the death of their abbot, some time towards the end of the sixteenth century, concerning the election of the new abbot. The difficulty lay in the fact that, at the time, the monastery included monks of many nationalities – Serbs, Greeks, Bulgarians and Russians. The Greeks proposed a Greek, because the monastery stands on Greek territory; the Serbs advocated a Serb, because Serbs had founded the monastery; the Bulgarians maintained it should be one of them, because they were the most numerous; while the Russians demanded that it should be a Russian, because the monastery received most of its funds from Russia. At the height of the dispute, one morning as they arrived for matins, the monks found the icon of the Three-Handed Virgin occupying the abbot's throne. Convinced that this must be a prank played by one of the inmates, they returned the icon to its original place. However, the following morning the same thing happened again. The church was then firmly locked at night and a guard placed around it to preclude any further interference. Even this precaution proved futile, as the next morning the icon was again found in the seat intended for the abbot. Then one of the Holy Mountain's most respected hermits, acknowledged by all to be of impeccable virtue, had a dream in which the Virgin appeared to him with instructions that peace and reconciliation must once more prevail at the monastery, for she would herself, henceforth, be its abbess. Since that time, the icon of the Three-Handed Virgin stays at the abbot's throne. Thus, until very recently, Chilandariou has been the only monastery without a formal abbot. This was changed by the decision of the Ecumenical Patriarch when he instituted a coenobitic system in the monastery, whereby the community was obliged to appoint a recognised leader. Even so, in Chilandariou, after every act of worship, the monks ritually bow in reverence and kiss the icon, as they would kiss the hand of the abbot in other monasteries.[4]

During the Russo-Japanese war of 1905, the Russian command requested that the icon of the Three-Handed Virgin be sent to them, in order to help the Orthodox army. Instead of the original, the council of monastery elders sent a copy. This was somewhat smaller in size and there were moreover on either side of the Virgin portraits of St Simeon and St Sava. It is maintained that the Russian army was successful (or at least that it suffered no defeats) whenever the icon was present, and that very shortly after its arrival at the front, peace was signed. Although the cult of the Three-Handed Virgin had existed in Russia as well, as far back as the time of Ivan the Terrible, the copy of the icon was later returned to the monastery where it can still be seen today on the right-hand side of the altar.

The second miracle, corroborated by witnesses, some of whom are still alive today, took place in 1945 when one of the frequent fires on Mount Athos swept through the forest surrounding Chilandariou. The fire spread rapidly, and was quite close to the monastery when the monks set out in procession, carrying the icon of the Three-Handed Virgin. Eyewitnesses swear that at this very moment a mighty wind started to blow from the opposite direction and pushed the fire away from the monastery buildings. It is a fact that one can still see the remains of the fire (charred tree-stumps and roots) very close to the south side of the walls surrounding Chilandariou.

The third case related by the monks of Chilandariou happened at about the same time, during the Greek civil war. The monastery was raided by number of bandits, said to be Markos partisans, but just as they were loading the stolen treasures onto their mules they were overcome by an inexplicable dread and ran away in fearful panic, leaving all their spoils behind. The surviving monks are convinced that they overheard the robbers speaking among themselves of a vision they had seen: a woman of superhuman height moving around the church, although she was invisible to the monks themselves.

The most recent miracle, documented in the monastery archives, occurred in 1992 in the town of Kragujevac in Serbia. A short-circuit in the wiring installations of a photographic studio, belonging to one Nebojša Nikolić, caused a fire to break out. Every single item of wooden furniture was burnt, every metal object melted out of all recognition, only a plastic apparatus for developing films and a cardboard calendar with a reproduction of the Three-Handed Virgin which had been left on top of it, remained intact. Photographs taken at the scene of the fire, showing the undamaged machine and the calendar appeared in the press and the monks, of course, enjoy showing these to visitors.[5]

In Russia, the cult of the Three-Handed Virgin was and remains unusually widespread.[6] When Serbia finally fell under Turkish domination, in the fifteenth century, most sources of income to Chilandariou dried up and the monastery found itself in dire financial straits. The monks were forced to go out on regular begging missions ('supplications') and to seek aid at the Russian court. The most generous patron was Tsar Ivan IV, better known as Ivan the Terrible, whose grandmother was Serbian. Many of the priceless gifts he donated are still kept in the monastery treasury. The Russian faithful had always been particularly sensitive to the monastic values of the Holy Mountain, as to anything that emanated from Athos, as though it had some special significance for them. For example, in Russia today the holiest of all icons is considered to be that of the Mother of God Iverskaya, which is in fact a copy of an icon known as Portaitissa from the once Georgian and now Greek monastery Iviron, an icon which found its way to Russia only as recently as 1655.[7] In a similar way and around the same time (c.1666), a monk by the name of Theophanes the Serb brought a copy of the Three-Handed Virgin from Athos to Russia and presented it to the Patriarch Nikon. The icon was placed in the Novojerusalimsky monastery in the vicinity of Moscow and the fame of its miraculous properties rapidly spread throughout Russia. Copies of this copy of the Chilandariou icon then proliferated in their thousands and the cult which had started in monastic and aristocratic circles pervaded all strata of believers.

In clerical circles generally, and monastic circles in particular, scholarly investigations into a monastery's past are understandably regarded with a certain disapproval. This especially applies in the case of such a deeply hallowed object as the icon of the Three-Handed Virgin. Nevertheless, it is incumbent on the scholar to present a summary of the known facts that research has so far established.

For example, it is well known that there was a church in Skoplje, built during the reign of King Milutin (i.e. some time between 1282 and 1321) which was dedicated to the Virgin Who Has Three Hands. There can also be no doubt that there was an icon of that name in the church, but nothing further is known about it. This is in obvious contradiction to the generally accepted belief that the icon was at that time resting in Chilandariou, where it had been brought by St Sava from Palestine at the beginning of the thirteenth century. In other words, this could be considered, at the very least, as reliable evidence that several replicas of the Virgin with Three Hands were in existence. It is also known that, in 1347, when the Serbian king Stefan Dušan was crowned Emperor (Tsar) of the Serbs and Greeks, the church of the Three-Handed One in Skoplje was elevated from a bishopric to an archbishopric. Sources are also quite definite that on the site of the church, which had been of modest dimensions, Dušan's heir Uroš the Infant later built a new church and dedicated it to the Holy Trinity.

What could be the significance of all this? The young Serbian academic, Srdjan Djurić proposes the following hypothesis: although the legend that St Sava brought the miraculous icon from Palestine to Chilandariou is colourful and widely accepted, it is actually of relatively recent origin.[8] If the original icon, associated with St John Damascene existed at all, then its trail fades and is lost somewhere between Palestine and Constantinople. The relics of St John Damascene were brought to Constantinople between 1187 and 1283, but no mention is made of the icon. However, it is quite possible that the Serbs came to hear of the legend and adopted the cult at the turn of the century (between the twelfth and thirteenth centuries), through Byzantium, or rather Constantinople.

Viewed from this perspective, it could be assumed that King Milutin built a church in Skoplje some time around the year 1300, dedicated to the Virgin With Three Hands, and had an appropriate icon painted for it. Djurić even thinks that Dušan commissioned a new icon which would 'resemble the image of the old one' specifically to mark the occasion of the church's elevation to the rank of Archbishopric. It is more than likely that during the reign of Uroš, Dušan's heir (1356–71), there was no longer such an icon in Skoplje, because the church of the Three-Handed Virgin had ceased to exist and on its old foundations there now stood a larger, new church of the Holy Trinity. It is therefore assumed that the icon of the Three-Handed Virgin in Chilandariou today is the one which caused the legend to arise and spread. It must have been brought to the Holy Mountain from Skoplje and here the crucial date, terminus post quem, could only be the year 1347 in which Dušan was crowned Tsar and the church attained the status of archbishopric.

If we follow this line of reasoning, we come to the next question: how and why was the icon taken from Skoplje to Chilandariou? When Tsar Dušan, together with his wife the Tsarina Jelena and their children, took refuge in Chilandariou from the plague which was raging towards the end of the year 1347, it is logical to suppose that he took the icon of the Three-Handed Virgin with him. He spent at least half a year there and when the danger from the plague subsided, he probably donated the icon to the monastery as a mark of gratitude. The belief that the icon also protects from the plague most likely dates from this time as well. It is interesting that a document in the monastery dating from 1804 tells of a monk who died of the plague in that year. The fraternity then took the icon of the Three-Handed One and carried it in procession round the monastery praying and invoking 'the intercessions of our little Abbess, the Virgin, to compel the deadly disease to disperse like clouds driven by the wind and to prevail upon it to desist and flee from this house'.[9]

If we accept this reconstruction of the icon's history, the question still remains as to why no record of it exists in any of the written medieval chronicles. Firstly, it is known that the monastic tradition has always and everywhere jealously guarded the cult of miraculous icons and surrounded it with a veil of enigmatic circumstances. Secondly, Tsar Stefan Dušan, although credited with bringing the Serbian medieval state to the greatest power in all its history, was not popular in clerical circles and was the only ruler of the Nemanjić dynasty not to be later canonised. He was never forgiven for bringing his wife to Mount Athos. His audacity in bringing a woman to the sacred land, the holiest of holies, has remained in the memories of generations and is still looked upon as an unforgivable sin. It is also known that the Tsar, during his stay at the monastery, insisted on collective monastic discipline, and maybe even laid down conditions regarding his future endowments to the monastery, all of which could not have been popular and has left a blight on his memory. Any or all of these could be valid reasons why the monastic records that have come down to us, are silent on the subject of Tsar Dušan's involvement in the acquisition of the famous icon for Chilandariou. It is, possibly, also the reason why they have chosen to support the more attractive version of its Syrian-Palestinian origins and the part played by St Sava, who was a deeply respected figure and the father of the Serbian Church, Serbian culture and spirituality in general.

Art historians and icon specialists are mostly in agreement that the icon of the Three-Handed Virgin in Chilandariou is not older than the fourteenth century.[10] The greatest authority in the field was the late Professor Vojislav Djurić. He soberly concludes:

> We are dealing here with a large processional icon, covered with a riza of silver and gold embellished with gifts, beneath which is concealed the Virgin with Christ painted in the style of an Odigitria (or Guardian of Travellers). The third hand is not painted but wrought in silver and, as legend has it, added on later. A frontal portrait of St Nicholas is depicted on the reverse side. It was the work of a very skilled artist from about the middle of the fourteenth century, during the reign of Tsar Dušan. It is probably a copy of an older, legendary icon. Rather serious and robust of form, the painting still bears the marks of the Palaeologus Renaissance style. It was the work of a Byzantine master.[11]

One other interesting feature, which does not necessarily contradict Professor Djurić's conclusion regarding the Byzantine origin of the iconograph, but does cast some doubt on it, is the fact that on either side of the icon there are inscriptions in Greek, but on the side depicting St Nicholas, among the Greek lettering a Slav, i.e. Serbian character has inadvertently been inserted, obviously in error.[12] If we allow for the possibility that the inscriptions on the icons were written by another hand, and not necessarily that of the painter, and that they could have been written at a considerably later date than the creation of the icon itself, Professor Djurić's conclusions remain valid. One should perhaps, all the same, keep an open mind as to the possibility that the painter of the icon and the writer of the inscription were one and the same person, possibly an iconographer of Slav descent, educated in Byzantium.

And here, finally, is a curious and fascinating fact. In the village of Karan, near Užice, on Serbian soil, in a little church known as the White Church of Karan, there is a fresco depicting a Three-Handed Virgin. This appears to be the only known example of this motif to be executed in the al fresco technique, albeit with substantial differences between it and the Chilandariou icon. The tricheirousa in the White Church of Karan – itself a work of outstanding quality – shows a very elongated, slender figure (220 cm in height) with a small Christ child held in her right hand. The left hand rests on her breast and, directly beneath it, there is a third hand of exactly the same size and character, identical in every way. No similarity whatsoever, therefore, with the Chilandariou icon. The little White Church is believed to have been built in the tenth century and it was certainly painted and embellished in the fourteenth century. Since it is now well established that the frescos themselves date from the fourth decade of the fourteenth century, it can be deduced that the painter of the Tricheirousa fresco must have known of the Three-Handed Virgin from hearsay or legendary tradition, but had never actually seen the Chilandariou icon. Here we have very persuasive evidence of how widespread and all-pervasive was the cult in medieval Serbia (and throughout Russia in the seventeenth and eighteenth centuries). The legendary nature of this icon was such that, frequently, even its name alone was enough to inspire the painting of the renowned image.

Following the scholars discussed, we have to conclude that the cult of the Virgin Who Has Three Hands must be older than the icon which is kept in the Chilandariou monastery on Mount Athos. The cult originated in the Middle East and arrived in the Balkans by way of Byzantium. It is probable that between the eighth and fourteenth centuries several such icons were circulating parallel with the cult, and that legend later fused them all into one. Even if the Chilandariou Virgin did not come into being until the fourteenth century, it has meant a great deal to the Serbs throughout the ages and it still does today. It symbolises a fusion of all their ancient beliefs: it is seen not only as their guardian, but as their protector from strife among themselves and conflicts with their neighbours, ever dispensing the blessings of peace and reconciliation.

Notes

1 *Povest o čudotvornoj ikoni Presvete Bogorodice 'Trojeručice'* (Manastir Hilandar, 1996), pp. 13–15.
2 *Вишнии покров над Афоном (или Сказани о святых чудотворнгх на Афоне прославившихся иконах* (Moscow, 1902), pp. 26–8.
3 *Povest o čudotvornoj ikoni*, pp. 18–19.
4 *Ibid.*, p. 23.
5 *Ibid.*, p. 26.

6 For a detailed account, see S. Petković, 'O kultu svetogorskih ikona u Rusiji', *Druga kazivanja o Svetoj Gori* (Belgrade, Prosveta, 1997), pp. 122–54.

7 Архимандрит Сергии, *Иверская святая и чудотворная икона вогоматери на Афоне и списки ее в Росии* (Moscow, 1879).

8 Srdjan Djurić, 'Hilandarska Bogorodica Trojeručica', *Kazivanja o Svetoj Gori* (Belgrade, Prosveta, 1995), pp. 100–13.

9 *Ibid.*, p. 111.

10 S. Petković, *Hilandar* (Belgrade, 1989), p. 35.

11 D. Bogdanović, V. Djurić and D. Medaković, *Hilandar* (Belgrade), p. 112.

12 Djurić, *op. cit.*, p. 109.

8
Hesychasm in the Balkans

Muriel Heppell

The word 'hesychasm' is derived from the Greek ἡσυχία (*bezmulvie* in Slavonic languages), for which there is no really satisfactory English equivalent; sometimes it is translated as 'quietude', or 'stillness', sometimes not translated at all. The word is used to denote a state of mental and emotional tranquillity: a mind freed from all distracting thoughts and images, which was considered to be a necessary preparation for the mystical contemplation of the Godhead (θεωρία).[1] The ultimate goal of this experience was a vision of the uncreated light seen by the disciples of Jesus during his Transfiguration on Mount Tabor;[2] according to hesychast theology, this uncreated light represented the energies of the Godhead, as distinct from the essence, which could never be seen by human eyes.[3]

The preliminary state of 'hesychia' could be achieved only by following a rigorous ascetic discipline: this included not only the usual monastic discipline of 'withdrawal from the world', and self-denial as regards food, sleep and conversation, but also a special method of praying, according to precise instructions as to breathing and posture;[4] and in addition the intensive use of the so-called 'Jesus Prayer': 'Lord Jesus Christ, have mercy upon me'.[5] It was thought that the constant oral and mental repetition of this prayer caused it to be 'drawn into the heart', which was regarded as the psychosomatic centre of the human personality. This was described by hesychasts as 'prayer of the heart', by which they meant not sincere, heart-felt prayer, but rather the continuous prayerful activity of the heart, which continued even when the mind and limbs were engaged in other activities.

This was the theory and practice which characterised the so-called 'hesychast movement' which developed in the Orthodox Church during the fourteenth century, and continued into the first decades of the fifteenth century. Neither the use of the Jesus Prayer, nor the more traditional type of ascetic discipline practised by the fourteenth-century hesychasts was new. Their basic monastic discipline can be traced back to the Desert Fathers of the fourth century, while the use of the Jesus Prayer is attested at least as early as the sixth century.[6] The discipline of 'guarding the mind', that is, excluding from it all intrusive thoughts or seductive images, is described in detail by the seventh-century writer St John Climacus in his work *The Ladder of Divine Ascent* (Κλιμαξθείαςάνόδου; in Slavonic *Lestvica*). This became a very popular work with the fourteenth-century hesychasts.[7] Another influential writer was the eleventh-century mystic Simeon the New Theologian (d. 1022). In fact, hesychasm, as it developed in the fourteenth century, was a typical renewal movement, drawing its inspiration from older traditions, but creating from them a new synthesis. This is especially true of the varied activities, practical and intellectual as well as spiritual, which characterised fourteenth-century hesychasm.

The origin of this movement is usually attributed to a Byzantine monk known as Gregory of Sinai (so-called because he was tonsured in St Catherine's Monastery on Mount Sinai). He spent the early part of his life travelling round the Byzantine world in search of spiritual enlightenment; in the course of his wanderings he visited Crete, where he learned the technique of 'prayer of the heart' from a monk named Arsenios. Afterwards he spent some years on Mount Athos, teaching this to other monks; his disciples included Bulgarian and Serbian monks, as well as Byzantine Greeks. In 1325 he left Mount Athos, because of the disturbances caused by Turkish pirates operating in the Aegean Sea, and sought refuge on the Balkan mainland, in a region known as the Paroria, described as 'the borderlands between the Bulgarians and the Greeks.' (Its exact whereabouts is unknown, but it must have been somewhere in the Strandzhija mountain range which straddles the present-day Bulgarian-Turkish frontier.) However, he soon returned to Mount Athos; but some ten years later, c.1335, he again fled to the Paroria to escape from the Turkish pirates. This time he stayed until his death, in 1346. His settlement in this remote part of Bulgaria marks the significant beginning of hesychasm in the Balkan peninsula.[8]

A key figure in this development was a Bulgarian monk known as St Teodosi of Turnovo. This designation is in fact misleading, since he had no permanent connection with Turnovo, then the capital of the so-called Second Bulgarian Empire.[9] Very little is known about his early life, except that he was tonsured in a monastery dedicated to St Nicholas in Vidin, in the far north-west of Bulgaria. During the period of his early monastic training there, he devoted considerable time to the study of *The Ladder of Divine Ascent*. Perhaps it was the influence of this work, stressing the importance for a monk to find the right spiritual father, which caused him to set out on a spiritual pilgrimage, shortly after the death of the abbot of St Nicholas – who had perhaps been his first teacher. This was the first of three long journeys which he was to make in the course of his life. During this first journey, he traversed the entire territory of Bulgaria, from the north-west to the south-east. This journey was not motivated simply by restlessness; Teodosi was consciously seeking for a teacher who could help him to advance in his spiritual life, and in doing this he was following a well-established tradition in Orthodox monasticism, in which the virtue of *stabilitas* was not considered as important as it was in the West. In the course of his journey, he visited a number of monasteries, including one in Turnovo. His biographer compares him to an industrious bee, going round different kinds of flowers, collecting their honey. Finally, he reached the Paroria, where Gregory of Sinai had already settled. He knew then that his search was over; he had found his teacher.

It seems that Gregory at once recognised the spiritual potential of the new arrival; possibly he experienced an intuitive awareness of a personality similar to his own. Certainly Teodosi's life, as he matured, was to follow a pattern very similar to that of his teacher. We are told that 'Gregory led him gradually along the path of spiritual progress... And having seen his divine zeal for the godly life, the great teacher omitted nothing, but taught him very thoroughly, and with great skill, instructing him not only in the ordinary ways of monastic life, which is the first stage, but also in the guarding of the mind, which gives protection from demons.' This passage from the Life of St Teodosi gives a good picture of the training of a young monk by his spiritual father, which was one of the characteristic features of Orthodox monasticism.

During this period of monastic apprenticeship, Teodosi made two visits to Turnovo, in order to ask for help from the ruler of Bulgaria, Tsar John Alexander, because the monks of the Paroria had suffered severely from raids by the bands of robbers who infested eastern Thrace at that time. Tsar John Alexander, a typical pious medieval ruler, responded generously to this request, so that the monks were able to build a tower in which they could take refuge, and also a church.

In 1346 Gregory of Sinai died, and the monks of the Paroria asked Teodosi to take his place as their leader and teacher; it is significant that the world 'abbot' is not used by this community, which was probably of the loose-knit type known as a *lavra*, in which the monks spent most of their time alone in their cells. However, Teodosi firmly refused this request, and actually left the Paroria, and set out on the second of his long journeys. This time he had a companion, a monk named Roman (sometimes also known as Romil). Together they visited Mount Athos, after which Roman returned to Bulgaria, while Teodosi went on to visit Salonica and Constantinople, where he stayed for a short time. Then he returned briefly to the Paroria, in order to visit the tomb of Gregory of Sinai. After this, he and Roman set out together once more and travelled to the town of Mesembria on the Black Sea coast, and stayed for a time in a local monastery; however, they soon left, when it was attacked by 'robbers and murderers'. It would seem that Teodosi was now engaged on another search, this time for a place where he could settle and follow the hesychast way of life in peace, and instruct others in it; once again, his biographer used the image of the industrious bee. He finally found the right place, with the help of Tsar John Alexander, who suggested that he should settle in a village not far from Turnovo called Kefalirevo, where there was already a small monastic community. This was to be Teodosi's home for the rest of his life.

Soon there was a community of some fifty monks at Kefalirevo, drawn there by Teodosi's reputation as a holy man, just as he himself had been drawn to the Paroria by the reputation of Gregory of Sinai. As well as spending time on the training and instruction of his monks, Teodosi also worked as a translator, translating from Greek into Slavonic texts which were important for those who wished to follow the hesychast way of life. The most important of these were a biography of Gregory of Sinai (written by one of his disciples named Kallistos), and also a selection of Gregory's own writings on the ascetic life, known as τα κεφαλαια πανμ ὠοφελίμα (*The Most Beneficial Chapters*). This work was later incorporated into an anthology known as the *Philocalia* (*Dobrotoljubie* in Slavonic), which is one of main sources for the study of hesychast spirituality.[10]

Like many holy men of the Middle Ages, Teodosi had a premonition that the end of his life was near; and it was this that caused him to set out on his third and final journey. Before he died, he wanted to see once more his old friend Kallistos (author of the Life of Gregory of Sinai), who was now patriarch of Constantinople; he also wanted to revisit the Church of the Holy Wisdom (Hagia Sophia) in Constantinople, 'the mother of the Orthodox churches'. First he wrote to Kallistos, asking whether it would be convenient for him to come on a visit; when Kallistos replied in the affirmative, he set out, accompanied by four of his disciples. After his arrival in Constantinople, he spent many hours in private conversation with Kallistos, and also in theological discussion with members of the permanent synod at Constantinople, which had been specially convened in his honour. But Teodosi had soon had enough of sight-seeing and meeting people. 'He liked best of all to live in holy silence', writes his biographer, 'and asked to be given an opportunity to continue in this way.' So the Patriarch arranged for him to go to the monastery of St Mamas, outside the city, where the hesychast way of life was strictly followed. And there he died, on 27 November (on the same date as his beloved teacher, Gregory of Sinai), probably in the year 1362.[11] Before he died he dictated a spiritual testament to the disciples who had accompanied him to Constantinople. It is a short text, but important because it is the only direct example of his teaching that has survived.

And when he [Teodosi] had reached a state of extreme weakness, and was about to depart to God, he summoned his disciples and spoke to them the following words: First of all, hold fast to the holy faith of the Church of the Apostles and Councils, and it its unshakeable precepts. Shun, as unfitting, the Bogomil[12] and Messalian heresies, and after that those of Barlaam, Akindynos, Gregory and Athanasios.[13] Believe those things which we have received from the beginning, without removing or adding anything; for this leads to blasphemy. This is what caused Akindynos to blaspheme, when he described Christ's glory, which at one time shone forth in a truly glorious and miraculous way, as something created. Likewise, keep the holy commandments. Hold fast to both these things; and a true Christian – by name, deed and repute – in addition roots out the love of self-will. Do not burden your life with possessions; practise fasting and self-denial, and so lull your passions. Subdue anger and all forms of bodily commotion, and [thus] drive away spiritual darkness. To speak briefly, this dries up all the moisture and sweetness of the flesh. He whose spiritual eye is clear sees himself in the manner of the pious David and overcomes the realms of evil, that is, the cunning inward thoughts of our hearts. Keep constantly and clearly before your eyes the remembrance of death and the Judgement of the Saviour, who will judge everyone and render to each according to his deeds. Have constantly and clearly before you the vision of God, as an activity of the mind; for this is a powerful weapon unswerving against all opposing forces. Above all, hold fast to love, the supreme virtue, with all your strength, for this is the fulfilment of all blessings. Make all strangers welcome; do not make *false* accusations, and avoid anger, rage, remembrance of wrongs and hatred; for these things darken the soul and estrange it from God.[14]

These may not have been the exact words spoken by Teodosi as he lay dying; the important point is that they represent what his disciples, or his biographer (or both), believed to be the quintessence of his teaching.[15]

In fact, this brief text, with its precise theological instruction, its perceptive psychological analysis, and its clear practical directives, manages to say a great deal in comparatively few words. It is interesting to note that it contains no reference to the Jesus prayer, or to the 'body language' – posture and breathing – that were part of the hesychast method of praying. By this time, and in this context, they were no doubt taken for granted as part of the hesychast training. In any case, the importance of these physical aspects of hesychasm have, in my opinion, been exaggerated by people with only superficial knowledge of the movement. Certain aspects of the testament, such as the reference to the 'clear spiritual eye', and the stress on the need to be constantly on one's guard against the 'cunning inward thoughts of our hearts', reflect the influence of *The Ladder of Divine Ascent*, which Teodosi had studied intensively as a young monk in Vidin.[16] This work contains many examples, some very vivid, of the way in which hidden, unconscious motives can restrict progress in the spiritual life. For example:

> I have seen many different plants of the virtues planted by those living in the world watered by vanity as if from an underground cesspool, made to shoot up by love of show, fertilised by praise; and yet they quickly withered when transplanted into desert soil.[17]

It is quite probable that the exposition and interpretation of the *Ladder* was a special feature of Teodosi's personal teaching.

Finally, the exhortation to love and forgiveness in the last sentence raises interesting questions. Had Teodosi found his monks deficient in these virtues? (Monastic literature has plenty of stories about quarrelsome monks!) Or was he trying to warn his disciples against too much self-centred absorption in their personal spiritual progress, which was one of the potential dangers of hesychasm?

The official Life (*Zhitie*) of St Teodosi (on which the foregoing pages are based) is an important source for the study of the hesychast movement in Bulgaria. Unfortunately, this work presents a series of textual problems which have so far prevented it from being fully utilised as an historical source.

First of all, the text of the *Zhitie* is preserved in only a single manuscript, included in the *Panegirik* of the Bulgarian monk known as Vladislav Grammatik. It was edited and published in 1904 by V. I. Zlatarski, in an academic publication not easily available outside Bulgaria;[18] as far as I know, it has not been reprinted. This text was translated into modern Bulgarian by V. S. Kiselkov, and published in 1926.[19] Again, as far as I know, it has not been translated into any other language.

More seriously, there are problems concerning the authorship of the text, which have given rise to doubts about the authenticity of at least part of its contents. According to the title of the single available manuscript, the *Zhitie* was written probably immediately after the death of Teodosi (c.1362), by Patriarch Kallistos of Constantinople.[20] He had known Teodosi well earlier in his life, when they were both disciples of Gregory of Sinai in the community of the Paroria; in fact, their relationship is an example of the type of close personal friendship which was characteristic of the hesychast movement. The strength of this particular friendship is attested by the fact that Teodosi travelled to Constantinople just before he died, in order to see Kallistos once more.

The Kallistan authorship of the *Zhitie* is supported by a number of first-person passages at the end of the text, in which the author refers to himself as the current patriarch of Constantinople,[21] recalls his earlier association with Teodosi,[22] and describes in detail Teodosi's final visit to Constantinople and his death there.[23] However, the *Zhitie* also contains information about the religious and political situation in Turnovo in the middle decades of the fourteenth century which Kallistos is unlikely to have known. Moreover there are internal inconsistencies in the text, notably in the portrayal of Teodosi himself: sometimes he is presented as a typical solitary or hermit, living in seclusion, withdrawn as far as possible from worldly affairs; sometimes just the opposite, as playing an active role in the stormy ecclesiastical conflicts of the time, and as being particularly active in opposing the Bogomil heresy, which then had many adherents in Bulgaria. There are also inconsistencies, at some points, between the *Zhitie* and the Life of Gregory of Sinai written by Patriarch Kallistos where there is no doubt about the authorship. Finally, there is the fact that there is no trace of the original Greek text, which must have existed at some point, if Kallistos was the author.

All these factors caused Kiselkov to reject the Kallistan authorship of the *Zhitie*. He considers that it was written by a Bulgarian monk, whom he tentatively identifies as Vladislav Grammatik; however, he thinks that one of the sources used by this writer was, probably, a short account of Teodosi's life written by Patriarch Kallistos shortly after Teodosi's death, based on material supplied by the four disciples who had accompanied Teodosi to Constantinople; hence the tradition of Kallistos as author preserved in the title of the *Zhitie*. Kiselkov presents his arguments in a long introduction to his modern Bulgarian translation of the *Zhitie*.[24]

Although I agree with much of Kiselkov's argumentation, I cannot accept his final conclusion. It is obvious that the *Zhitie* as it stands is a compilation. However, I would suggest that the core of this text is *a* short biography of Teodosi written by Kallistos, based on his own personal recollections, as well as material supplied by Teodosi's disciples. In this text Teodosi was portrayed consistently, and, I believe, correctly, as a reclusive type of hesychast, who avoided contact with the outside world as far as possible, and concentrated, first on his own spiritual development, then on teaching and training his disciples. I think that it was probably written at the request of Teodosi's disciples in Constantinople, who thought that the patriarch's authorship would give the work added prestige, and help to stimulate further interest in hesychasm in Bulgaria. This text was then immediately translated into Bulgarian by one of the disciples. Since it was intended for use in Bulgaria, there was no need to copy the Greek text, hence its disappearance need not cause problems for modern scholars.

Later, this short Life written by Kallistos came into the hands of a Bulgarian copyist who thought it would be improved by the addition of material from his own knowledge of events in Turnovo during the later part of Teodosi's life. Perhaps he felt that the work needed livening up a bit, since it must be admitted that Kallistos writes in a somewhat sedate and formal style. This copyist had, in fact, a considerable amount of authentic material at his disposal; and we should be grateful to him for preserving it. But he is inclined to stress sensational elements, and uses his material uncritically. For example, he confuses the monk Teodosi with a contemporary of the same name who was then Patriarch of Turnovo. It was this Teodosi who energetically opposed the Bogomil heretics, and not the hesychast Teodosi. This would explain the ambivalent portrait of Teodosi in the extant *Zhitie*.

I advance this conclusion tentatively, since I have not yet tested it by a detailed examination of the style of the different parts of the work. This would have to include an investigation of the various metaphors derived from apiculture, which are a characteristic feature of the *Zhitie*. If it could be established that Patriarch Kallistos had at some earlier stage in his life been in charge of the monastic beehives, this would certainly be a persuasive argument in favour of his authorship.

I have discussed the *Zhitie* of St Teodosi in some detail, because I consider that further work on this text is urgently needed. The first task would be to try to establish a reliable text on the basis of the single manuscript so far available. Then it would be necessary to clarify the authorship of the different parts of the text, and estimate their authenticity and historical value. Finally, the text would need to be annotated, and then translated into languages other than Bulgarian, in order to make it accessible to a wider readership. Only then will the *Zhitie* be able to take its rightful place as a major source, both for hesychasm in the Balkans and for the history of the Second Bulgarian Empire during the middle decades of the fourteenth century.

I shall now deal briefly with some other aspects of hesychsm, as it developed in the Balkans. The earliest hesychasts, both in the Byzantine Empire and in the Balkan peninsula, were solitaries, that is, monks who spent most of their time alone, engaged in prayer and meditation, having only occasional contact with their fellow monks, and little, if any, with the outside world. As I have already indicated, I consider that Teodosi's life followed this pattern, apart from his journeys. However, fairly early in the history of the hesychast movement, we find examples of hesychast monks who abandoned the solitary, and even the monastic life, in order to pursue an active career in ecclesiastical administration. For example, there were four hesychast patriarchs of Constantinople in the second half of the fourteenth century: Isidore (1347); Kallistos I (1350–3, and 1355–63); Philotheus (1353–4 and 1362–76); and Kallistos II. Hesychasts whose life followed this pattern have been described as 'urban hesychasts'.[25] All these patriarchs were Byzantine Greeks. In the Balkans, the two most notable examples of this type of hesychast were Patriarch Evtimi (Euthymius, the last patriarch of Bulgaria, 1375–94); and Kiprian (Bulgarian by birth), Metropolitan of Kiev and All Russia from 1390 to 1406. Both these monks were disciples of St Teodosi, and, incidentally, close personal friends. Both of them, in addition to being very energetic and successful administrators, were also writers. However, with regard to Metropolitan Kiprian, we do know that throughout his busy life in Muscovy (Russia), and in spite of the ecclesiastical conflicts in which he sometimes embroiled himself, he never ceased to be a practising hesychast; that is clear from a passage in a contemporary Russian chronicle, under the year 1406:

Kiprian fell ill in the village of Golyanishchevo, which was part of the metropolitan's property. He often liked to go there, and write books with his own hand, and to consecrate bishops, practise pure prayer, read sacred books, and meditate on the remembrance of death, the Last Judgement, and the torments of the unrighteous.[26]

This evidence, of course, relates to only one man; but I think it is reasonable to assume that other 'urban hesychasts' made similar short periods of withdrawal, of the kind that would now be described as retreats.

Another feature of Balkan hesychasm was the stimulus it provided for translating Greek texts into Slavonic. The primary reason for this was to make the teaching of Gregory of Sinai available to monks who did not know Greek. St Teodosi translated both the *Life of Gregory of Sinai* by Patriarch Kallistos, and parts of Gregory's teaching on the spiritual and ascetic life known as *The Most Beneficial Chapters*. There were also new translations of *The Ladder of Divine Ascent*, including one made by Metropolitan Kiprian; and numerous other translations. In fact, the large number of extant copies of this work, dating from the late fourteenth and early fifteenth centuries is one of the indications of the spread of hesychasm beyond Bulgaria, into Serbia and the Romanian principalities of Moldavia and Wallachia. One translator is mentioned by name in the *Zhitie*: a monk named Dionisi, a member of the community at Kefalirevo, of whom it is said: 'he received from God the gift of translating from Greek into the Slav languages with marvellous skill... and he translated many books, thus enriching the condition of the Church.'[27]

So far, most of what has been said has related to the development of hesychasm in Bulgaria. However, it is clear that it spread beyond Bulgaria to Serbia and the Romanian principalities, though the evidence for this is more fragmentary. One indicator, already mentioned, is the number of extant manuscripts of texts connected with hesychasm, such as *The Ladder of Divine Ascent*, the writings of Gregory of Sinai, and, in Serbia, translations of the works of Gregory Palamas, the leading theologian of hesychasm. Gregory Camblak, the nephew of Metropolitan Kiprian, spent some years in Serbia in the early fifteenth century, as abbot of the Pantocrator Monastery in Dečani. In one of his numerous writings, he comments on the high standard of monastic life in Serbia, where, he noted, many of the monks were practising hesychasts.[28] A key figure in the development of hesychasm in Serbia was, in all probability, the monk Roman (Romil), who accompanied St Teodosi on the second of his long journeys. He eventually settled in Serbia.[29] Another Bulgarian, a generation younger, called Konstantin Kostenetski (sometimes known as Constantine the Philosopher), also eventually settled in Serbia, though he had his early monastic training in Bulgaria, under a disciple of Patriarch Evtimi. He wrote a long and detailed biography of the Serbian ruler Despot Stefan Lazarević, and also a treatise on the orthographic reforms associated with Patriarch Evtimi.[30]

In the Romanian principalities, it is mainly the survival of manuscripts that indicates the spread of hesychasm. There are several of *The Ladder of Divine Ascent*, including one copied by the Moldavian scribe Gavril Uric, famous for his exquisite calligraphy. Later, the Neamţ Monastery in Moldavia, founded by a monk named Nicodemus who had spent his early years on Mount Athos, became an important spiritual centre.[31]

It is important to note that the influence of hesychasm was confined to the Orthodox areas of the Balkans. Although there were contemplative and mystical movements in monastic circles in the Catholic Church, certain aspects of Byzantine hesychasm, notably the vision of the uncreated light on Mount Tabor as representing the 'energies' of God, were regarded with suspicion in the West, and even considered heretical.

It might be argued that the hesychast movement in the Balkans is no more than a historical curiosity, of interest only to a few ecclesiastical historians. I should like to conclude by suggesting that this is not a valid interpretation. Religious movements which certainly have some elements in common with hesychasm (though not necessarily sharing its theology) can be found in Western Christianity in the later Middle Ages. They also occur in more recent times, and in other faiths, especially those originating in Asia. Hence some knowledge of fourteenth-century hesychasm is useful both for ecumenical discussion among Christians, and in Inter-Faith Dialogue, which currently has a high profile. The hesychast movement which flourished in the Byzantine Empire and the Balkan peninsula in the fourteenth century also has some relevance for our own times.

Notes

1 For a further exposition of the concept of 'hesychia', see the Introduction by Bishop Kallistos Ware to the English translation of *The Ladder of Divine Ascent*, trans. Colm Luibheid and Norman Russell, in *Classics of Western Spirituality*, London, SPCK, 1982, pp. 50–3.
2 Matthew 17: vv. 1–9.

3   John Meyendorff, *Byzantine Theology*, New York, 1983 (2nd edition), pp. 77–8.
4   See Kallistos Ware, 'The Jesus Prayer in Gregory of Sinai', *Eastern Churches Review*, 4 (1972), pp. 14–16.
5   See *The Jesus Prayer by a Monk of the Eastern Church* (Revised translation, with a foreword by Kallistos Ware), New York, St Vladimir's Seminary Press, 1987.
6   *The Jesus Prayer*, pp. 36–7.
7   This is attested by the large number of manuscripts of *The Ladder of Divine Ascent*, dating from the fourteenth and early fifteenth centuries; see Dimitrije Bogdanović, *Jovan Lestvičnik u vizantijskoj i staroj srpskoj književnosti*, Belgrade, Byzantine Institute, Special Publications, II, 1968, pp. 204–7. (I understand that more of these manuscripts have been discovered since this work was published.)
8   For a detailed account of the early life of Gregory of Sinai, see Kallistos Ware, 'The Jesus Prayer in Gregory of Sinai', pp. 4–7. (This is based on a Life of Gregory by Kallistos I, Patriarch of Constantinople, 1350–3 and 1355–63.)
9   For a brief account of the emergence and subsequent history of the Second Bulgarian Empire, see D. Obolensky, *The Byzantine Commonwealth* (London, 1971), pp. 219 and 243–7; and M. Heppell, 'The Hesychast Movement in Bulgaria. The Turnovo School and its relations with Constantinople', *Eastern Churches Review*, 7 (1975), pp. 10–11.
10 An English translation of the original Greek texts is now in progress by G. E. H. Palmer, the late P. E. Sherrard and Kallistos Ware. Four volumes have already been published; vol. V is in progress.
11 This account of Teodosi's life is based on his *Zhitie* (vita), written shortly after his death: 'Zhitie i zhizn' prepodobnago otsa našego Teodosiia', ed. V. I. Zlatarski in *Sbornik za narodni umotvoreniia i knižina*, vol. XX, Sofia, 1904, pp. 1–41. Hereafter, *Zhitie*.
12 The Bogomil heresy was an uncompromising form of dualism which had many adherents in Bulgaria in the tenth century. In spite of rigorous measures taken against it, it flourished once more in Bulgaria in the fourteenth century.
13 Barlaam, Akindynos and Gregory all opposed both the hesychast theologian Gregory Palamas (1296–1359), on the distinction between the 'essence' and 'energies' of the Godhead. Barlaam's views were condemned by a specially convened church council in 1341, and those of Gregory Palamas officially approved by two subsequent councils in 1347 and 1351. (See John Meyendorff, *Byzantine Theology*, pp. 76–8; and *The Jesus Prayer by a Monk of the Eastern Church*, p. 59.)
14 *Zhitie*, pp. 33–4.
15 M. Heppell, 'The Hesychast Movement in Bulgaria', p. 15.
16 *Zhitie*, p. 11.
17 *The Ladder of Divine Ascent*, p. 82.
18 See note 11.
19 See V. S. Kiselkov, *Zhitie na sv. Teodosii Turnovski kato istoricheski pametnik* (Sofia, 1926).
20 See *Zhitie*, p. 9.
21 *Ibid.*, p. 31.
22 *Ibid.*
23 *Ibid.*, pp. 31–4.

24 Kiselkov, *op. cit.*

25 See Kallistos Ware, 'The Jesus Prayer in St Gregory of Sinai', p. 6.

26 'Patriarshaia ili Nikonovskaia Letopis", *PSRL*, 11 (1962, reprint), p. 195.

27 *Zhitie*, p. 18.

28 See M. Heppell, *The Ecclesiastical Career of Gregory Camblak*, London, 1979, p. 32. In his account of the translation of the relics of St Pareskeva from Bulgaria to Serbia, Gregory refers to monks in Serbia as 'standing in the heaven of stillness'. (See E. Kaluzhniacki, *Werke des Patriarchen von Bulgarien Euthymius, 1375–1393*, p. 436, Vienna, 1901).

29 See Heppell, *op. cit.*, p. 30. There is a Life of St Romil, by a monk named Peter. See F. Halkin, 'Une érémite des Balkans au XIV siècle. La Vie grecque inédite de St Romylos,' *Byzantion*, 31 (1961), pp. 149–87.

30 See Heppell, 'The Hesychast Movement in Bulgaria', pp. 18–19.

31 See E. Turdeanu, *La Littérature bulgare du XIV siècle et sa diffusion dans les pays roumains*, Travaux publiés par l'Institut d'Etudes slaves (Paris, 1947).

Part III
Islam

9
Islam in the Balkans: the Bosnian Case

Alexander Lopasic

It was long believed that the process of Islamicisation in Bosnia owed much to the rapid conversion to Islam of many members of the nobility and adherents of the 'Bosnian Church'. There are two sources which support this belief. One was the report of the papal legate to Bosnia, Nikola of Modruš. He tried to explain the speedy and catastrophic end of the Bosnian kingdom in 1463 as the result of 'heretical and treasonable behaviour of Manichaeans' who betrayed their king and the Church to the Ottomans and accepted Islam in large numbers in return for pre-servation of their position and property. The importance of Nikola of Modruš has been discussed at some length by the Bosnian historian, Srečko Džaja.[1] The fact was that it was Bishop Modruš who persuaded the last unfortunate ruler of Bosnia, Stjepan Tomašević, to refuse the payment of vassalship to Mehmed II, and accept instead an alliance with Hungary and the help of Christian powers. This advice had disastrous consequences for Bosnia, because of the weak response of the Christian side.

At the same time there is a report of the re-conversion of 12 000 members of the 'Bosnian Church', and a group of about 40 members who found refuge at the court of the Knez Stjepan Vukčić-Kosača. This report is among the documents of the Roman Curia.[2]

Obviously some pockets of the 'Bosnian Church' survived in north-east Bosnia and Hercegovina. Also Ottoman defters studied by the Turk-ish historian of Bosnian descent, Tayyib Okić, give us a number of details about the 'Krstjani' communities after the Ottoman conquest.[3] The idea of 'Bogomil betrayal' was later accepted by the early twentieth-century Croatian historian, Ćiro Truhelka, who put forward the idea of 'mass conversion to Islam in order to preserve positions and property', as it supported the idea that the Muslim nobility was of Christian background.[4] That idea was challenged for the first time by the Serbian historian, Vasa Čubrilović, who pointed out, in 1935, the varied origin of the Muslim Bosnian nobility. According to him only a small number of that nobility belonged to the old Bosnian aristocracy; the rest were recruits of the 'Devshirme' system who achieved responsible positions, as Janissaries or palace officials, in the Ottoman Empire. Otherwise they were Sipahis from Asia Minor who settled in Bosnia. A small number, however, were of Croatian, Dalmatian, Serbian or Hungarian origins; they entered Bosnia as a result of wars and changing frontiers.[5]

Two more important studies followed, one by the Bosnian historian, Behija Zlatar, who, between 1976 and 1978, investigated the background of 48 noble Muslim families with 180 individuals from the sixteenth century, and the other by her fellow-historian, Ahmed Aličić who studied the background of 3116 cavalrymen from Bosnia who participated at the fateful battle at Mohacs in 1526.[6] Of 48 family names, twelve were certainly of the pre-Muslim Christian nobility, nine, very probably seven, were certainly in Bosnia in the sixteenth century; four were of south Slav origin, but definitely from outside Bosnia, five were non-Slav in origin, and the remaining eleven were unknown or possibly non-Slav in origin.[7]

Aličić published the translation of the battle-order of Sipahis from Bosnia (Yoklama defteri), excluding Hercegovina, which is often treated separately, giving us a reckoning of 16 senior commanders, 213 members of their retinue, 1040 members of the Sandjak Bey's following and 338 ordinary Sipahis. According to an analysis of 1600 soldiers from Bosnia, of whom sixteen senior commanders were certainly Muslims of the second or third generation, 52 of the Sipahis were Christians, mostly Vlachs, with small Timars (about 15 per cent of the total) fighting for the Sultan, while twelve of the Muslim Sipahis were of Christian origin. Seventeen had their fathers fighting in the Ottoman army, and were mostly of Christian descent. Of the members of the retinue 72 per cent were from Bosnia with no further details available, 17.5 per cent were from Croatia and the remaining 10 per cent were from Europe and Asia, including 31 Albanians and twelve Hungarians. All Croats and Hungarians were described as Gulams (Slaves) which meant that they were recently converted to Islam, while the Bosnian 72 per cent were described as Muslims, which means at least second generation, and some of them must have been from the rest of the Ottoman Balkans or even Asia.

It is interesting to compare this with the different figures for Muslims and Christians from the Ottoman defters, as they give us a more accurate idea about the Islamicisation in Bosnia which was a gradual and, indeed, slow process. The Ottoman defters from 1468/9, dating from only six years after the Ottoman conquest, and published by Nedim Filipović, show us a moderate number of Muslim households and a small Muslim population. The figures are: 37 125 Christian family households, 8770 Christian single households, 147 Christian widow households, 322 Muslim family households, of which 263 were in villages and 68 in market places and towns. About half of the total population lived as ordinary peasants in the countryside, the other half in towns and market places, often enjoying the privileged position of craftsmen, miners, traders or Vlachs.[8] These defters indicate a small number of Muslims who were either landed nobility or officials or soldiers of town-garrisons. Some of the small towns like Foča, Rogatica or Visoko in Eastern Bosnia became Muslim before the Ottoman conquest of Bosnia in 1463; Rogatica and Visoko also had *tekkes* (retreats, chapels and charitable hostelries of the Muslim Sufi orders).[9]

The defter of 1484 shows us a different picture of steady Islamicisation over eighteen years, from 322 Muslim households in 1468 to 4200 in 1485. The figures are: 30 552 Christian family households, 2443 Christian single households, and 48 Christian widow households, against 4134 Muslim family households and 1064 Muslim single households.[10] The newly established Muslim ruling strata included both local Muslims and Muslims from outside Bosnia, but still included some Christian families. It seems that the changes included movements of non-Islamicised members of families to more compact Christian areas, leaving more urbanised and Islamicised centres with newly converted Muslims and Muslims who came with the Ottoman administration and the army.

In this way the future religious map of Bosnia started to take shape. Even if cases of pressure were known, the actual use of force is denied by important Christian sources of the time (e.g. Benedikt Kuripešić).

One important factor contributing towards Islamicisation was the depopulation of large areas of Bosnia and Hercegovina due to war and the flight of the Christian population to both Croatia and Hungary. One of the consequences of the depopulation was the movement of Vlachs who accompanied the Ottoman army as auxiliary forces and settled down in abandoned villages, encouraged by the local Ottoman administration which was very interested in repopulation of devastated areas. In due course they became a sedentary population but still kept some sheep and *katuni* (shepherd's pens). In the defter of 1468/69 we already have information about Vlach settlements in Hercegovina (Vilayet

Herzeg) where 4616 Vlach family households and 998 Vlach single households were recorded. In Vilayet Pavli (Pavlovići) 448 Vlach family households and 130 Vlach single households were counted.[11]

Defters from 1485 and 1489 mention the further arrival of Vlach colonists in north and central Bosnia (Maglaj and Visoko), though sometimes as ordinary rayah with no special Vlach privileges. Vlachs were already known in the days of Christian Bosnia as traders and entrepreneurs, moving from one part of the country to another. As many Vlachs in due course became Muslim, so 'Vlachisation' and Islamicisation developed hand in hand.

It is also of some interest to look at the background of early converts to Islam and for this Tayyib Okić provides some important information. The defter of 1468 describes fourteen Sipahis as new Muslims (Muslimi nev) and one who became converted after surrendering the fortress of Samobor. Also Christian names of brothers and fathers are mentioned for the next eleven Sipahis clearly indicating their Christian origin. The majority were Bosnians, but six were Hungarians, three Albanians, four Vlachs and two Germans. The last two must have been Saxons, who came to Bosnia as miners and enjoyed different privileges under the Ottomans too. Many of them would in due course become Muslims.

Some of these new Muslims became soldiers of fortress garrisons which included Sipahis from other parts of the Ottoman Balkans (Macedonia, Bulgaria or Serbia).[12] Fortresses played an important role in Islamicisation and here Zvornik and Srebrenica are good examples. Zvornik became the main Ottoman fortress performing an important strategic role, though many were in Hungary. It was known as 'the key post on the river Drina'. It became the centre of the Sandjak of Zvornik and part of Bosnian Pashalik until the end of Ottoman rule in 1878.

The battle at Mohacs in 1526 and the ensuing stabilisation of political and economic conditions certainly contributed to the growth of towns and an increase of population became clear in defters from 1533 and 1548. The increase of population was also due to migrations from other parts of the Ottoman Empire, primarily other parts of the Balkans. Good examples of this are towns such as Zvornik, Donja Tuzla, Gornja Tuzla, Jasenica of Bjeljina, where Islamicisation and urbanisation went hand in hand.[13] In 1533, the Nahiya of Zvornik had fifteen villages with 640 households, of which 230, or about 36 per cent, were Muslim. The town of Zvornik had 113 Muslim and only seventeen Christian households.[14] In 1548 the Nahiya had 1077 households of which 623 were Muslim (about 57 per cent). Srebrenica, as an old mining centre, doubled its Muslim population between 1533 and 1548, becoming, in the end, an important Ottoman-Oriental town, still preserving, however, its orginal mining character.[15]

Towns became not only centres of the Ottoman administration and the garrison cities, but also foci of the Islamic way of life. The practice of creating towns was always strongly supported by Islam, as only with a mosque, a market and a public bath can the requirements of the Muslim faith be satisfied. In a number of cases the establishment of a town began with the building of a mosque. A number of towns in fifteenth- and sixteenth-century Bosnia were built around a mosque, notably the future capital, Sarajevo, which is a Muslim creation. Some of the most important towns built in such a way were Zvornik, Srebrenica (the Muslim part), Foča, Višegrad, Travnik, Prozor, Doboj, Bjeljina, Glamoč or Kulen Vakuf. The idea of a 'kasaba' (enclosed Muslim quarters and towns) was connected with the creation of a mosque and a 'ferman' confirming 'kasaba' status. Some of the mosques were built in the name of the Sultan: e.g. Sarajevo's in the name of Mehmed II, also the mosque at Zvornik; the mosque of Foča, Rogatica, Višegrad or Travnik in the name of Bajazid II (1481–1512).[16] The creation of 'Waqfs' (religious endowments) was another factor in the establishment and development of towns, supporting Ottoman imperial policies.

It is not, therefore, surprising that the defters of 1520–35 show us a rather different picture for Bosnia and Herzegovina. Bosnia then had 16935 Muslim households as opposed to 19619 Christian households. Herzegovina, as a separate Sandjak, shows 7077 Muslim and 9588 Christian households. There is no doubt that in Herzegovina a decisive role was played by large numbers of Vlach immigrants, who settled in abandoned villages and accepted Islam.[17]

In this period, Sarajevo had 1024 Muslim households and, by comparison, Skopje had 630 Muslim, 200 Christian and twelve Jewish households. Salonica, on the other hand, contained 1229 Muslim, 989 Christian and 2645 Jewish households.

Great Ottoman victories, including the Hungarian disaster at Mohacs, also encouraged Islamicisation and the population of north-east Bosnia, studied by Adem Handžić, became one-third Muslim by 1533. Some of this Muslim population moved to newly conquered Slavonia in the latter half of the sixteenth century. It should be added too, that the important centres of the Catholic Church in north-east Bosnia, like the monasteries of Zvornik and Gornja and Donja Tuzla, became isolated from their flock, which fled towards the west to escape war. The remaining population surrounding the monasteries became Muslim in the second half of the sixteenth century.[18]

Closely connected to town and mosque foundation was the establishment of Muslim tekkes. Isa-bey, a leading early Ottoman administrator of Bosnia, founded a tekke, leaving for its upkeep some land and a few mills. Eventually it became a children's home, an inn and a public kitchen.[19] An important example of a tekke which eventually became a Bosnian order is one in the village of Orlovići (between Vlasenica and Zvornik) founded by Sheikh Hamza, known also as Hindi Hamza Dede in 1519. Sheikh Hamza was originally a Sipahi who left the army in order to found a tekke and spread Islam in a non-Muslim area. At that time there was only one mosque in the fortress of Zvornik. Fortresses were not only centres of the Ottoman army but also of early Islam. The tekke of Orlovići became famous and developed into the Bosnian Hamzevi order, named after its founder. The order was influential not only in Bosnia but in neighbouring Hungary and even in Istanbul. The order was connected with the assassination of Mehmed Pasha Sokolović (Sokolë) in 1579; he was known as a bitter opponent of the influential Hamzevi order.[20]

The urbanisation and Islamicisation of Bosnia and the rest of the Balkans were certainly closely connected. One should not forget, however, that part of the urban population always remained Christian, Jewish or in some cases Gypsy. The Muslim quarter was controlled by an Imam, the Christian mostly by an Orthodox priest. Since the Patriarch was in Istanbul, in the Ottoman administrative system Christians were represented by the Orthodox Church, as the two main enemies of the Ottomans were Austria and Venice, both Catholic powers. Catholics had certain difficulties, even though Apostolic visitors were never refused permission to visit their flocks. Some of their reports represent an important source of information about the position of Catholics in Bosnia or, indeed, in Ottoman Europe.

Religious leaders performed religious and other communal functions and represented their communities in the Ottoman state. Such communities were known as Millets (from the word *millä*, meaning nation, people or religion). Each of them had its own courts, schools, places of worship and welfare. The system was stratified and the Muslim Millet, representing the religion of the state, had the highest position. On the one hand the Millet system preserved the local autonomy and cultural and religious identity of different ethnic groups in the Balkans but, on the other, it perpetuated differences and therefore prevented the creation of one coherent Ottoman society. The reforms of the nineteenth and twentieth centuries proved how difficult modernisation was under such conditions.

An important local institution which crossed the religious divide was that of the guilds, which played a significant role in all Ottoman cities. Different trades and crafts, some newly established under the Ottomans and often related to the army, brought together different religious groups. They followed a prescribed set of rules and were an important source of state revenue. There were many guilds in Bosnia but Sarajevo remained the economic centre, where various disputes were settled by the courts.[21]

One particularly important Ottoman institution was the 'Devshirme', the 'boy tribute', or 'tribute in blood', a periodical levy of Christian children for the service of the Ottoman state, primarily the Sultan himself. They were trained to serve as Janissaries (the new army, an elite force of 10 000–12 800 men in the second half of the sixteenth century), or as court or administrative officials. Literally the term means 'collection of Dhimma-children', children of Christians (Dhimma) who lived under Muslim protection. The levy was primarily concentrated on village children, and excluded children from urban centres, families with one son, orphans, shepherds or sons of local dignitaries. The number recruited varied between 1000 and 3000 per year, and even larger figures were known in war years. The primary task of the Devshirme was to provide a reservoir of physically and mentally fit young men for the service of the Sultan. At the same time the Devshirme institution guaranteed the regular conversion of young men who, in due course, not only became part of the Ottoman élite but were often dedicated Muslims as well.

The Devshirme has provoked considerable controversy among scholars of Balkan origin as it has often been presented as a conscious attempt by the Ottomans to deprive the Christian communities of their most gifted individuals. Historians sympathetic to the Ottomans have emphasised the opportunities for promotion and advancement in the ranks of the Osmanli élite. Some critics, like Basilike Papoulia, see it as primarily a 'forcible separation of children from their Christian parents and inclusion into Ottoman Muslim society'.[22] The role of Bosnia in the Devshirme system is particularly interesting as it reflects the territory's special position as a frontier province and as an important source of soldiers for the Ottoman army. As early as 1515 we have a record of the recruitment of 1000 young men for Janissary service. Such recruitments were connected with a legend, according to which, during his campaign in Bosnia, Mehmed II developed such a liking for the large number of Bosnians who willingly accepted Islam, that he permitted them to join the Janissaries, whether they were Muslim or not. It is true that in

Bosnia the Devshirme included both Christian and Muslim children. One should also mention the particular division of Devshirme, at the end of the sixteenth and in the seventeenth centuries, according to which the levy distinguished between those who were circumcised but ignorant of the Turkish language, called 'Potur' and those who spoke Turkish (Türkesmis). It seems that the background of the story was related to the refusal of some Bosnians to enter the Devshirme system. There are even cases of exemption from the Devshirme in the sixteenth and seventeenth centuries.[23]

Another Bosnian institution which contributed to Islamicisation was the Military Frontier, organised by the Ottomans as a response to the Austrian Military Frontier in the sixteenth century and known in Ottoman documents as *Serhat* (a Persian-Arabic word meaning 'frontier region' or 'frontier'); *Serhatlije*, on the other hand, meant 'frontier soldier'.[24] Commanders of such frontier units were known as *Kapetani* (Captains) and their provinces were called *Kapetanije*. Their pay was better than that of ordinary soldiers and they enjoyed a number of other privileges and a higher status. The number of *Kapetanije* increased from 20 at the end of the sixteenth century to 39 at the end of the eighteenth. Their main function was to guard and patrol the frontier which included a number of fortresses, and their pay was related to service in such fortresses. It is important to emphasise that in due course their positions became hereditary, and Captains became a real Bosnian frontier aristocracy, accumulating over the years influence, power and property, and becoming in the end leaders of Bosnian Muslim society. Some of these frontier Captains developed true local dynasties which acquired considerable autonomy in dealing with their Christian neighbours. The families of two of these Captains, the Čengići and the Rizvanbegovići, were among the famous names which were still prominent and active as late as the nineteenth century. *Kapetanije*, found in Ottoman documents under this name, indicating both their Bosnian origins and the Austrian model, were a unique feudal institution known only in Bosnia because of its position as a frontier province of the empire.[25] Around 1830, there were some 24 000 soldiers serving in frontier forces under 39 Captains in an area extending from Bihać in the north to Počitelj in the south.[26] The first *Kapetanije* were recorded as early as 1558: for example, Gradiška, Krupa, Bihać and Gabela.[27]

The appointment of Captains closely concerned local Bosnians. The Captains often established contacts with their relatives on the other side of the frontier. This is well documented in the voluminous correspondence which developed between commanders on both sides, often written in Croatian, using the Bosančica script, a simplified form of the Cyrillic script particularly popular among the Muslims (used even by Muslim women in correspondence and poetry). Some of the Captains were the sons or grandsons of converts, and some originated from noble Christian families.

The frontier organisation evolved with time. It began as a springboard of Ottoman conquest, inspired by the frontier spirit in Asia Minor. Later developments led to a new type of frontier, with the function of inter-action between two opposing worlds, Muslim and Christian, rather than its original purpose of being a focus of invasion and further conquest in the West. An important aspect of the frontier society was the develop-ment of extensive trade links, as well as smuggling: of cattle, sheep, leather, tobacco, coffee, sugar and salt, which were much cheaper on the Ottoman side. Venetian Dalmatia and the Republic of Ragusa also partcipated in these lucrative operations. But there were hazards as well, such as quarantine against plague, which often endangered the Balkan world.

The Military Frontier in Bosnia had many similarities with the situation in Spain, where a similar frontier society between Muslim and Christians developed between the twelfth and fourteenth centuries. Facing each other and fighting each other in the name of Islam and Christianity eventually brought people closer, not only for the exchange of material goods, but also in spiritual attributes and ideas. Wars and destruction led eventually to human relations and a new tolerance.[28]

In the last century the frontier Captains started to play an important new political role, as members of special councils of Ayans (councils of local dignitaries) advising the Vizier of Bosnia, representing the Sublime Porte. Owing to the fact that Captains passed on their function, either from father to son or to a near relative, the combination of their political and economic power made them into semi-independent rulers of their *Kapetanija*. This independent policy led them in the early nineteenth century to oppose vigorously the policies of reform (Tanzimat) and the loss of their influence, and to open revolt by one of the Captains, Husein Beg Gradaščević, who was eventually beaten by pro-government forces. The ultimate consequence was the abolition of the *Kapetanijas* in 1835 and the introduction of sweeping government reforms. This explains the lukewarm reaction of the Bosnian Muslim élite towards the Berlin Congress in 1878 and the Austrian occupation of Bosnia.[29] On the other hand, the Ayans represented an interesting experiment in Ottoman urban organisation.

In the countryside Islam developed much more slowly, as may be clearly seen in various defters, especially the one of 1520/30 during the reign of Suleiman the Magnificent. In that often-quoted defter, there are 833 000 Christian households as opposed to 195 000 Muslim ones: a proportion of 1:4, whereas Muslim households clearly predominated in towns. Of some importance here is the frequently quoted document from 1469 from a village of Dušina (Kreševo Nahija, an important mining town of Christian Bosnia, west of Sarajevo). This concerns a land dispute between Muslim converts and a local Ayan (Muslim dignitary). The peasant converts claimed the land as their *Baština* (patrimony), a Slav term recognised by the Ottomans ever since. The document further demonstrates that Islamicisation was spreading in the countryside five years after the Ottoman conquest, around Kreševo, which became an Ottoman fortress shortly after the conquest.[30]

Islamicisation in general, and in villages in particular, increased in the 1530s and 1540s, as may be seen in the defters. In the second half of the sixteenth century, we have a different picture. The great Ottoman defeats at Lepanto (1571) and Sisak (1593), as well as ensuing economic difficulties, slowed the rate of Islamicisation. Here again Adem Handžić provides details for north-east Bosnia (Donja and Gornja Tuzla). In the former, the Muslim population around 1533 represented 12.5 per cent of the total and between 1533 and 1548 30 per cent, but between 1548 and 1600 only 2.5 per cent more.[31]

Gornja Tuzla shows different figures and the increase went from 44.6 per cent in 1548 to 54.7 per cent in 1600. These changes apply to both the towns and the villages. The Christian population of the villages fell by 15 per cent, which may be explained by the increased migration of the Christian village population into towns.[32] The Ottoman administration did not encourage large-scale movement of the peasantry to towns, and imposed special taxes on the owners of unused land in order to make rural–urban migration more difficult. In 1608, for instance, taxes and fines were imposed and villagers had to wait for ten years to be recognised as citizens (šeherlije).[33]

The reasons for the Islamicisation of the peasantry lie in a combination of different factors, including the systematic increase of urbanisation, the accumulation of wealth by some trading families and the general well-being of Ottoman society in the sixteenth century. In addition we should mention the large-scale participation of Muslims in wars, often as recruits of the Devshirme system and the continuation of family ties after Islamicisation, resulting from the strong south Slav patrilineal family system. The original south Slav kinship system was based on large extended families known as *rod*, which continued their function when families, or parts of them, became Muslim. There were some changes of emphasis; for example, families with typically long genealogies going back seven or more generations became rare, and families of three generations were considered old. One reason for this was connected to conversion to Islam a few generations back, coupled with frequent unwillingness to admit Christian forefathers. Muslim families continued to keep close ties with their neighbours, manifested through mutual help during the harvest or other agricultural activities or the regular exchange of visits and invitations for Bayram, weddings, etc.[34]

To sum up, we see among Bosnian Muslims two emerging groups: on the one hand the Muslim élite, consisting of landed and military nobility, high officials and clergy, who practised class endogamy, intermarrying with similar groups elsewhere in the Empire; and on the other, peasants and craftsmen who preserved a number of pre-Islamic institutions and customs such as exogamy, marriage by capture, the cult of the patron saint, godparenthood, belief in St George and St Elias (the traditional patron saints of cattle), and belief in witches, demons, the evil-eye and so on.

Because the Bosnian Muslims were mostly local people who accepted Islam and its institutions at different times over a period of 400 years, Islam remained more or less intact after the Ottoman departure in 1878. These 400 years had resulted in the creation of a real Muslim identity based on a civilisation combining Islamic and pre-Islamic elements.

The period of Austrian rule was characterised by the preservation of the status quo between the three religious groups. It introduced improved connections with Vienna, Budapest and the rest of Central Europe by road and rail. Modern administration and education also brought Bosnia closer to the rest of Europe. Muslim leaders used a number of tactics in order to preserve their old privileges and the new Austrian administration moved very carefully as they did not wish to upset the existing balance between the three religious groups. The Austrians studied the French and Russian experience in North Africa and Central Asia, and applied some of their conclusions to Bosnia. The first Austrian census of 1879 showed a population of 448 000 Muslims (39 per cent of the total population), 496 000 Orthodox (42 per cent) and 209 000 Catholics (18.5 per cent) and 14 000 Jews.

The Austrian administration recognised the Muslim religious hierarchy and the first Bosnian parliament formed in 1910 was organised strictly according to religious denomination. The same principle was applied to the town councils. In this period also the first political parties were formed following religious affiliaton. These were the Muslim National Organisation (Muslimanska Narodna Organizacija) and the Croat Muslim National Party (Hrvatska Muslimanska Narodna Stranka). These two parties later agreed to join in a United Muslim Organisation.[35]

During the Austrian period, between 1882 and 1903, the administrator Benjamin Kallay tried to foster the idea of a Bosnian nation, which, under the name of Bosniaks, received support from some groups, particularly the more conservative Muslims, and still has some importance today, as it is supported by one of the present Muslim parties, known as MBO, Muslimanska Bošnjačka Organizacija. Kallay's experiment failed, as at that time both Serb and Croat ethnic and political organisation was developing. But this period saw an emphasis on interdenominational balance, still supported by many Bosnian Muslim leaders today.[36]

Perhaps the most important result of Bosnia becoming part of the newly created state of the kingdom of the Serbs, Croats and Slovenes (after 1928, Yugoslavia), was the formation of the Muslim political party known as JMO (Jugoslavenska Muslimanska Organizacija). The party was founded to represent the interests of all Muslims of Bosnia irrespective of their social or economic background. JMO was a regional party, demonstrating that Muslims had their own social and cultural identity, despite considerable pressure, expropriation of large landed properties, and the chicanery of the new local administration in its efforts to destroy Muslim local influence and its economic base in both towns and villages.[37] The Yugoslav state took the view that Muslims were really Slavs who had abandoned their original Christian faith and become Muslims, as well as being traitors to their own people. In spite of that, the JMO had some success as it joined different coalition governments which needed its support, and also maintained a careful balance between the central goverment in Belgrade, mounting Serb-Croat differences and its own interests. The popular leader of the JMO was Mehmed Spaho, who proved himself an able tactician and champion of Muslim interests in a number of political battles in the Yugoslav parliament.

The first years of the new state were particularly difficult for the Muslims: their property was stolen or destroyed, and people intimidated, beaten up, or even killed by organised groups of their Serbian political opponents. There were a few serious clashes with the local gendarmerie in which a number of people were injured or killed. During the Second World War, the Muslims in Yugoslavia found themselves in a very precarious position. At that time, Bosnia became a battlefield for three different, imported political movements: 1. The Ustasha movement from Croatia, which tried to mobilise Muslims, offering them special status in the Croatian state, where Muslims became an important minority representing 12 per cent of the population in 1941 (717 000)[38]; 2. The Communists, who began to form partisan units after the German attack on the Soviet Union in June 1941, and tried to attract the Muslims on to their side, even organising special units consisting of Muslims only; 3. The Chetniks, Royalist Serbs, who at first adopted a very hostile attitude to the Muslims, killing about 8000 between August 1941 and February 1942.[39] Later they changed their policies and offered Muslims a place in their political programmes. Of the three main ethnic-religious groups in Bosnia, the Muslims suffered the greatest losses. According to B. Kočović, about 86 000 Muslims were killed.[40]

Because of the large-scale destruction of villages, particularly in eastern Bosnia and the Sandjak, the Muslim population lost confidence in the Croatian Pavelić regime and tried to find help elsewhere. This led to a particularly bizarre story according to which Muslims received support from the Waffen-SS and its leader Heinrich Himmler, who had been impressed by the Muslim troops fighting in the ranks of the Austrian army in the First World War. Himmler promised protection to Muslims in return for joining a special SS division (13th SS, called Handjar) which was trained in France to fight the partisans in Bosnia. However, after hearing that they would be sent to northern Bosnia instead of eastern Bosnia, they rebelled. The division was crushed and the remnants sent to northern Bosnia where many of its members either went over to the partisans or just fled home.[41] In 1943 and 1944, the Cazin region in western Bosnia became an autonomous area under the partisan deserter, Huska Miljković, who organised a private army and was supplied by both sides, until his assassination in the latter part of 1944, which ended another bizarre Bosnian story.[42]

In Socialist Yugoslavia, the Muslims were again an important issue. They were recognised as a religious community, and were allowed to claim the status of 'Non-Declared Yugoslavs'[43] until 1961, when they were given permission to call themselves 'ethnic Muslims'. In 1964 they were granted the right to self-determination and in 1971 they were recognised as a separate nationality. These developments provoked considerable polemics in the Yugoslav press, particularly in Serbia. The reasons for their recognition, which the Bosnian Muslims pursued vigorously, involved a number of factors, including their precarious experience during the Second World War. Every republic of Socialist Yugoslavia represented one ethnic group, so Bosnia came to represent the Bosnian Muslims, who had by then become its largest ethnic group. The region should not be underestimated; Bosnia was the core republic of Yugoslavia, and the centre of the Yugoslav armament industry. The impact of Tito's death made itself felt rather late in Bosnia, which was the last of the Yugoslav republics to introduce changes in the political structure, including the multi-party system and eventually separation from the rest of Yugoslavia. In the 1980s Sarajevo witnessed one of the most spectacular political trials of Communist Yugoslavia: a Sarajevo lawyer, Alija Izetbegović, now President of Bosnia, along with twelve Muslim intellectuals, was charged with 'hostile and counter-revolutionary activities'. In reality the regime was afraid that Bosnian Muslims might establish political links with other Muslim countries, thus offering them a chance to voice their own ideas and claims. Toward the end of 1988, the sentences of Izetbegović and his associates were suspended. The regime hoped to create a more tolerant atmosphere, but the press, partiularly in Serbia, continued to print anti-Muslim slogans and propaganda.[44]

In August 1991 Bosnia was declared a democratic, independent state of three nationalities, Muslims, Serbs and Croats, and the other ethnic groups living there. According to the census of that year, the Muslims represented the majority of the population: 43.7 per cent of the total population, with 31.3 per cent Serbs, and 17.30 per cent Croats. Other groups numbered 7.7 per cent. The official ethnic map of Bosnia shows the concentrated presence of Muslims in eastern and north-eastern Bosnia, with the important enclaves of Bihać, Cazin and Bosanska Kladuša in the north-west, as well as Bosanski Brod and Bosanski Šamac, where Muslims were mixed with Croats, in the north-east.

Many Bosnian Muslims feared that the introduction of a multi-party system would upset the precarious balance between the three main communities. The result of the first elections confirmed these fears, as they followed lines of loyalty based on ethno-religious affiliation. The Communists lost their importance. The election results also made it clear that the Muslim Party, SDA, Stranka Demokratske Akcije (Party of Democratic Action), had won the majority of seats, but not enough to form a government. They had to look for allies, and turned to the HDZ (Croat Democratic Union) with 44 seats. The situation started to deteriorate because of the war in Slovenia and neighbouring Croatia, which resulted in substantial military forces of the Yugoslav army moving into Bosnia and becoming concentrated there. The Muslim leadership tried to avoid confrontation with Serbia and the Yugoslav army, offering a loose confederation with Serbia and Montenegro which was rejected by the SDS, Srpska Demokratska Stranka (Serb Democratic Party), which had followed a hard line from the beginning. Various attempts by the European Union countries, including a well-thought-out plan by the Portuguese mediator Jose Cutileiro, failed. In the end, a referendum was agreed for 29 February and 1 March 1992. The Muslims and Croats together won 63 per cent of all the votes. The SDS proclaimed a boycott of the referendum and put up barricades around Sarajevo. There were several clashes, followed by unsuccessful truce arrangements by the European Union and the eventual intervention of the Yugoslav army. From April 1992 Bosnia became a battlefield of dramatic proportions with a huge number of casualties (some 200 000 killed) and a massive number of refugees dispersed throughout the world.

The post-Communist world order has begun very badly for arguably the oldest Muslim community in Europe, which now depends on outside help and the goodwill of her neighbours. The dramatic and cruel war in Bosnia exposed the Bosnian Muslims to 'ethnic cleansing', mass extermination, rape and uncertainty about their future.[45]

The Bosnian case is certainly tragic as it touches the question of basic human rights for a community of two million which has for 500 years tried to secure its place and survival in the most unstable part of Europe. However, after a terrible ordeal, Bosnian Muslims feel now more confident in their own abilities, as, after all, they have survived and can look to the future. Bosnia has established many international contacts, including membership of the Islamic World Conference. On the other hand, as a part of south-eastern Europe, it has its place in Europe, where it has belonged for several centuries. After all, Bosnia can exist only in a balanced system between two conflicting protagonists, Croats and Serbs, representing a force of reason, tolerance and co-operation, qualities which Bosnian Muslims have acquired in the course of their long and difficult history.

Notes

My preoccupation with Bosnia is long-standing and I would like to thank three individuals in particular for help and advice on a number of issues, such as Ottoman and Christian sources on Bosnia, the different aspects of the 'Bosnian Church', and relations between the three communities in Bosnia. These are: Emeritus Professor V. L. Ménage, formerly Professor of Turkish and Head of the Department of Near and Middle East Studies in the School of Oriental and African Studies, University of London, the late Dr. H. Šabanović, of the University of Sarajevo, and Dr. S. Džaja of the University of Munich.

1   S. Džaja, *Konfesionalitat und Nationalitat Bosnien und der Herzegowina* (1984), in particular pp. 25–7, 34, 231–2 (Latin original of the text).

2   *Ibid.,* p. 26.
3   T. Okić, 'Les Chrétiens de Bosnie d'après des documentas turcs inédits', *Südost Forschungen,* 1960, pp. 108–33.
4   Č. Truhelka, 'Historička podloga agrarnog pitanja u Bosni', *Glasnik Zemaljskog Muzeja u Sarajevu,* no. 27, 1915, p. 124.
5   V. Čubrilović, 'Poreklo muslimanskog plemstva u Bosni', *Jugoslovenski historijski časopis,* 1935, no. 1, pp. 388–402.
6   A. Aličić, 'Popis bosanske vojske pred bitkom na Mohaču 1526 g', *Prilozi za Orientalnu Filologiju,* 25, 1975, pp. 171–202. (English summary, pp. 201–2.)
7   B. Zlatar, 'O nekim muslimanskim porodicama u Bosni u XV i XVI st.', *Prilozi Instituta za istoriju,* 14–15, 1978, pp. 81–139.
8   S. Džaja, 'Die "Bosniche Kirche" und das Islamisierungsproblem Bosniens und der Herzegowina in den Forschungen nach dem Zweiten Weltkrieg', *Beitrage sur Kenntnis Sudosteuropas,* 28, 1978, pp. 71–2.
9   D. Čehajić, 'Derviški redovi u jugoslovenskim zemljana, sa posebnim osvrtom na Bosnu i Hercegovinu', *Orijentalni Institut u Sarajevu,* Posebno izdanje 14, 1986, p. 21.
10 Džaja, *op. cit.,* 1978, pp. 74–5.
11 *Ibid.,* pp. 75–6.
12 Okić, *op. cit.,* pp. 108–33.
13 A. Handžić, 'O islamizaciji u sjeverno-istočnoj Bosni u XV i XVI vijeku, 1978', *Prilozi za Orientalnu Filologiju,* 1966/67, pp. 28–9.
14 *Ibid.,* p. 31.
15 *Ibid.,* p. 32.
16 A. Handžić, 'O formiranju velikih gradskih naselja u Bosni u XVI st.', *Prilozi za Orientalnu Filologiju,* no. 25, 1975/77, pp 135–67; 'O gradskom stanovništvu u Bosni u XVI st.', *Prilozi za Orientalnu Filologiju,* 1978/79, pp. 247–55; 'O ulozi derviša u formiranju gradskih naselja u Bosni u XV st.', *ibid.,* no. 34, 1981, pp. 169–78.
17 O. L. Barkan, 'Essai sur les données statistiques des Registres de recensement dans l'Empire Ottoman aux XVe et XVIe siècles', *Journal of the Economic and Social History of the Orient,* 1, 1958, tables 6 and 7, pp. 31–3, 35.
18 Handžić, *op. cit.,* 1966/7, p. 9.
19 Handžić, *op. cit.,* 1981, pp. 171–2.
20 For more details and sources see: A. Lopasic, 'Islamisation of the Balkans with special reference to Bosnia', *Journal of Islamic Studies* (Oxford), 5, no. 2, 1994, pp. 169–70.
21 H. Kreševljaković, 'Esnafi u Bosni i Hercegovini', *Godišnjak Istorijskog društva BiH,* Sarajevo, 1949, pp. 169–205.
22 B. D. Papoulia, 'Ursprung und Wegen der "Knabenlese" im osmanischen Reich', *Süd-Ost Europaische Arbeiten,* 59, 1963, pp. 116 passim.
23 V. L. Ménage. 'Devshirme', *Encyclopaedia of Islam* (2nd edn), 1961, pp. 210–13; and 'Some Notes on the Devshirme', *Bulletin of the School of Oriental and African Studies,* University of London, no. 29, part 1, 1966, pp. 64–78; K. Binswangen, 'Untersuchungen zum Status der Nichtmuslime im Osmanischen Reich des 16 J.', *Beiträge zur Kenntnis südost Europas und des nahen Orients,* no. 23, 1977, pp 354–65.
24 M. Škaljić, *Turcizmi u srpsko-hrvatskom jeziku* (Sarajevo, 1966), p. 560.

2 H. Kreševljaković, 'Kapetanije u Bosni i Hercegovini', *Naučno društvo Bosne i Hercegovine*, no. 5, 1954, p. 7.
26 *Ibid.*, p. 31.
27 Kissling, 'Betrachtungen über Grenztradition und Grenzorganisation der Osmanen', *Scientia* (Milan), 11–12, ser. VII, 1969, pp. 1–10.
28 Lopasic, 1997, paper on 'Islam and the building of the European House' at British Society for Middle Eastern Studies meeting, Oxford, July 1997.
29 A. Sučeska, 'Ajani – Prilozi izučavanja lokalne vlasti u našim zemljama za vrijeme Turaka', *Naučno društvo SR BiH*, 22, 1978, pp. 156–66, 173–4.
30 G. Elezović, *Turski spomenici*, 1940, I, p. 76; H. Šabanović, 'Turski dokumenti o Bosni iz druge polovice XV st.', *Istorijski pravni zbornik*, Sarajevo, 2, 1949, p. 182.
31 Handžić, *op. cit.*, 1966/67, pp. 42–3.
32 *Ibid.*, p. 43.
33 Truhelka, *Historička podloga agrarnog pitanja u Bosni* (Sarajevo, 1915), pp. 30–1, 81–2.
34 For the kinship system of Bosnia, see: A. Lopasic, 'The "Turks" of Bosnia', *Research Papers: Muslims in Europe*, Birmingham, no. 16, 1982, pp. 12–23.
35 F. Hauptmann, (ed.), *Borba Muslimana Bosne i Hercegovine za vjersku i Vakufsku-Mearifsku autonomiju* (Sarajevo, 1967), pp. 24–5; R. J. Donia, *Islam under the Double Eagle. The Muslims of Bosnia and Hercegovina, 1878–1914*, East European Monographs, LXXVII (New York, 1981), pp. 170–80.
36 Donia, *op. cit.*, pp. 10–17; 50–5; 160–6.
37 A. Purivatra, *Jugoslavenska Muslimanska Organizacija* (Sarajevo, 1974), pp. 46–55, 111–14, 421–2.
38 F. Wiener, (ed.), *Partisanen-Kampf am Balkan, Truppen-dienst Taschen-bücher*, 26, Vienna, 1976, p. 98.
39 E. Redžić, *Muslimansko autonomaštvo i 13. SS-Divizija* (Sarajevo, 1987), p. 107.
40 B. Kočović, *Žrtve II Svjetskog rata u Jugoslaviji* (London, 1985).
41 L. Hory, and M. Broszat, *Der Kroatische Ustascha-Staat 1941–1945* (Stuttgart, 1965) 2nd edn, pp. 154–62.
42 In the same area, in the 1980s, the Muslim politician and entrepreneur, Fikret Abdić, created the largest Muslim agricultural enterprise, which eventually collapsed. During the most recent conflict, Abdić became the 'president' of a self-proclaimed Bihać autonomous republic, maintaining a balance between Croats and Serbs. In the end, that collapsed as well and Abdić moved to Rijeka.
43 A category in the census for individuals who did not fit into one of the available ethnic groups, or did not wish to describe themselves in that way.
44 For the Islamic Declaration and the trial of Dr Izetbegović, see: 'The Trial of Muslim Intellectuals in Sarajevo: The Islamic Declaration', *South Slav Journal*, no. 1, 1983, pp. 55–89; nos 1–2, 1985, pp. 94–7; S. F. Ramet, 'Primordial Ethnicity or Modern Nationalism: the Case of Yugoslavia's Muslims Reconsidered', *South Slav Journal*, vol. 13, no. 1–2, 1990, pp. 1–20.
45 It is perhaps worth mentioning the fact that the old idea of 'komšiluk', neighbourhood, so important to Muslims and intercommunal relationships, manifested itself in many ways between these groups during the siege of Sarajevo, where a number of Serbs remained, sharing with their Muslim and Croat neighbours the hardships and dangers of the besieged city.

10
Muslim Communities in Romania: Presence and Continuity

Jennifer Scarce

It is a well-known and understandable feature of the long period of Ottoman Turkish rule in Central and Eastern Europe that distinctive Muslim communities existed in Albania, Bulgaria, Greece, former Yugoslavia and, to a certain extent, Romania. What is less well-known, however, is that the presence of Muslims in Romania long precedes the Ottoman conquest, as it began with the immigration of Tatars from the ninth century onwards. Romania's position under Ottoman rule was unique, even privileged, and quite different from that of other Balkan territories.[1] Romania was never under direct rule, as the two provinces of Moldavia and Wallachia were treated as autonomous Christian provinces with their own native princes who, providing that they kept the peace and paid tribute, could manage their own affairs without the supervisory presence of an Ottoman governor. The much-disputed province of Transylvania was only under Ottoman supervision from the sixteenth to the seventeenth centuries, otherwise it was under Hungarian control. The only areas of Romania directly ruled by the Ottoman Turks were the Black Sea region of Dobruja from the late fifteenth century to the nineteenth century and the Banat in the west which was ruled from 1552 to 1699 by a series of Ottoman governors. As a result, Romania's Muslim community was always concentrated in the Dobruja region and was also ethnically distinct from the Romanians who formed the majority of the population.

Muslim presence was early in Dobruja, which was under nominal Byzantine control after the break-up of the Roman Empire but was threatened continuously from the ninth century by invasions by Turkic peoples. Geographically, Romania was accessible to immigrants from the east who passed through the Crimea and the Caucasus and moved down to south Moldavia and Dobruja by land and over the Black Sea.

This strategic vulnerability was continually exploited by Turco-Tatar peoples such as the Pechenegs, who dominated the land between the Dnieper and the Danube in the late ninth century and the Uzes, who arrived in the eleventh century and reached as far south as Thessalonika. The most important, however, were the Kipchaks or Tatars, who settled in significant numbers. The Tatars were originally part of the Mongol armies who had devastated Iran and neighbouring parts of the Middle East in the thirteenth century. After moving west and dominating the Crimea, they invaded the Dobruja in 1241. By the fourteenth century, there were significant communities in south Moldavia between the Prut and Dniester rivers and in Dobruja down to Babadag. This predominantly Tatar population was further enriched by immigrants of Anatolian Turkish origin. Michael VIII Paleologus (1261–82) encouraged two Seljuk Turkish leaders, Issedin and Sara Saltuq, to settle with their followers in the Dobruja, offering them lands in return for defending the frontier. They arrived between 1262 and 1264 and settled around Babadag where, long after his death, Sara Saltuq's tomb was a place of Muslim pilgrimage. The Ottoman Turks in the fifteenth century were therefore comparatively late arrivals in Romania. They concentrated on the Dobruja for several reasons. Strategically, the Dobruja was a convenient base for routes to the Crimea and points further north such as Russia and Poland, and also as an observation post for Moldavia and Wallachia. Economically, a good climate and fertile soil ensured a steady food supply for the Ottoman Empire. The Dobruja was also well-situated for the development of trade as total control of the Black Sea gave the Ottomans a commercial monopoly. The distinct presence of Muslims provided by the Tatar and Anatolian settlers was also an incentive to direct Ottoman administration. Sultan Mehmet II, after a long succession of campaigns, subdued the Dobruja in 1462. In 1484, the Dobruja, the Danube Delta and its environs, were formed into a single province with boundaries extending from Dniester in the north to Nicopolis in the south. By the sixteenth century, the vilayet of Silistra was created with an Ottoman governor resident either at Silistra or Babadag. The Dobruja remained Ottoman within these boundaries until the Treaty of Berlin in 1878 which ratified the transfer of the province to the two independent Romanian principalities of Moldavia and Wallachia.[2]

The pattern of conquest and migration created an ethnically complex Muslim population which continued to be shaped by political events up to the twentieth century. The original communities of Tatars and Turks who had settled in the Dobruja since the ninth century were enlarged by further immigration. During the period of Ottoman rule, Turks steadily moved into key towns such as Babadag, Constanta, Silistra and Tulcea. They were followed by more settlements of Tatars. In the early sixteenth century, Crimean Tatars settled in Karasu (present-day Medjidia). Conflict between Turkey and Romania in the eighteenth and nineteenth centuries resulted in the loss of Tatar lands in Crimea and Bessarabia. Nogai Tatars in 1783 and Bessarabian Tatars in 1812 therefore migrated to the Dobruja. The Treaty of Berlin of 1878 again affected the Muslim population as many Tatars and Turks emigrated to Turkey. This pattern was repeated in the twentieth century, when both Tatars and Turks again emigrated to Turkey after the end of the Second World War.

The result of all these changes is a permanent Muslim minority circumscribed in both number and location. Accurate population figures are currently difficult to estimate: a round total of c.40 000 Muslims concentrated in the Dobruja was given in the census of 1983, but it is possible that these numbers have declined through further emigration. Since the revolution of 1989, the majority group is Tatar of Nogai and Altaic origin, concentrated in the town of Medjidia, but also in Constanta, Hirsova, Isaacea, Babadag and Mangalia and in about fifty villages. Turks are very much a minority, based mainly at Tulcea, supplemented by immigrants from Turnu Severin, Ada Kaleh and Timisoara, descendants of communities of the time when the Banat region was under direct Ottoman rule. Tatars and Turks have inter-married, so a typical 'Romanian' Muslim may well, for example, have a Tatar father whose ancestors migrated from the Crimea or the Caucasus and a Turkish mother from Tulcea.[3]

Romania's Muslims continue to observe the beliefs and rituals of Islam with varying degrees of commitment and piety, which has contributed another faith to the traditionally varied and complex religious pattern of Romania. They are too small in number to pose any challenge to the dominant Romanian Orthodox Christianity and make no efforts at conversion. Since 1878, they have become well-practised in accommodation to the requirements of the Romanian state. Their position after 1878 was favourable, since the Romanian government had declared that all citizens had equal rights, regardless of race and religion, and provided state support for the maintenance of mosques and Muslim training institutes, and subsidised the teaching of Arabic to Muslim children. These rights were further recognised and defined in the constitution of 1923 and the Religious Cult Law of 1928. A state-approved religious infrastructure was therefore in place well before Romania passed into Communist control, as a People's Republic in 1947. Within the context of an authoritarian atheist state, steps were taken to incorporate and regulate all religious practices. As with the Patriarch of the Orthodox faith, the Mufti, head of the Muslim community, was a member of the Grand National Assembly, while some state funds were allocated to supplement voluntary contributions towards the expenses of personnel and institutions. Islam effectively had and continues to have equal status with other religions.

Romanian Muslims have managed to achieve a realistic accommodation with the state for various reasons. Numerically, they were too small a minority to cause any effective opposition. They do, however, belong to a universal religion practised by affluent Arab states which were politically and economically important to Romania; any discriminatory treatment could provoke unwelcome repercussions. Islam is also, when compared with Orthodoxy, a relatively simple and unstructured religion which has neither an ascending hierarchy of clergy, sacraments, nor elaborate public rituals. It may therefore be observed discreetly when necessary for survival.

Within this broad framework it is of considerable interest to analyse the structures and beliefs of Romanian Muslims. They are Sunni Muslims of the Hanafi branch of Islamic law and as such are closest to Turkish practice, which is natural in the context of their deep-rooted contacts with the Ottoman Empire. This influence has also survived in religious terminology, which is Turkish as well as Arabic. The community is administered by the Mufti and his staff based in Constanta, who also appoint and supervise the work of about 120 Imams located in 50 towns and villages. There is also an Islamic centre in Bucharest. Certain fundamental beliefs are obligatory for all Muslims: Tawhid – the single nature of Allah; Risalah – the unique role of the Prophet Muhammad as an intermediary between Allah and mankind; and Akhirah – the existence of life after death. Complementary to these beliefs are the five essential duties of Islam which Romanian Muslims managed to accommodate in various ways. The first duty, the Shahadah, is a straight declaration of faith: 'There is no God but Allah and Muhammad is the Prophet of Allah', which can be simply and discreetly recited and does not require conspicuous initiation rites. The second duty, Namaz or prayer, requires Muslims to pray five times a day: at dawn, noon, late afternoon, after sunset and at night before sleep. This prayer system is clearly designed as a defining framework for a Muslim's daily life and other activities must be planned around it. In practice, flexible compromises were and are made in Romania, as in all modern secularised societies. Here, traditional Muslim practice is still helpful, as, apart from the noon prayer, the remaining four can take place at home or anywhere. Times of prayer can also be adjusted if necessary. Again, while it is desirable to perform the noon prayer daily in a mosque, attendance is usually only expected at the Friday noon prayers which are led by the community's Imam. It has therefore been possible for Romanian Muslims to make some effort to integrate prayer into a daily work routine by grouping prayers together at a convenient time or deciding to omit some; it is very much a matter of personal decision. Attendance at Friday noon prayer has been more difficult as it interrupts a working day and also depends on access to a nearby mosque. By current estimates, the number of working mosques is static, even dwindling, and apart from Bucharest, is naturally concentrated in the Dobruja. The main mosque in Constanta, the Hunkar Camii, is a conspicuous flamboyant structure built in 1910 by a Romanian architect, Gheorghe Constantinescu, on the site of an eighteenth-century mosque.[4] It is an attractive building, stylistically inspired by Ottoman Turkish mosque architecture, with a central dome, pointed minaret, and a colonnaded interior courtyard. Construction and decoration combine modern and traditional techniques, where marble facings and Turkish ceramic tiles are grated on to a reinforced concrete foundation. The mosque is open daily because it features as a tourist attraction, visited by passengers from the cruise ships which regularly anchor at the port of Constanta, and also by individuals browsing around the town. The Muftiate recognises this function and provides the services of its staff as guides, in return for a modest admission charge. In this mosque, communal Friday noon prayer is attended by about 100 Muslims. These are inevitably older, mainly retired men. Romanian Muslim women who could pray at the back of the mosque or on an upper balcony, or at home, make the latter choice if indeed they are religious. Apart from the Hunkar Camii, there are four small mosques in Constanta whose opening hours are variable. There are also old mosques ranging from the late fifteenth through to the mid-nineteenth centuries in Babadag, Isaacea, Mangalia, Medjidia and Tulcea.

The three remaining duties require a more communal form of involvement. They are: Zakat, charity; the Ramazan fast; and Haj pilgrimage. Charity is interpreted as a Muslim's obligation to the community which may be expressed in many ways, depending on an individual's means. Wealthy Muslims may donate funds for the foundation of mosques and associated religious institutions, often establishing Waqf trusts for the ongoing administration of the charitable bequest. A portion of a Muslim's income may be taken as a tithe towards communal social work. At a very modest level, a purchase of materials for cleaning the local mosque is an example of charity. In Romania, Muslim charity functions on an informal basis, directed towards social work. In Constanta, the Muftiate's staff would receive food left over from wedding parties and Bayram celebrations and dispense it to the poor, would regularly visit old people and, in some cases, fund their accommodation.

In Ramazan, the ninth month of the Islamic calendar, fasting takes place from dawn to dusk and is regarded as a Muslim's renewal of commitment to Allah. As Ramazan is calculated according to a lunar calendar, it is a seasonably variable ritual independent of the secular year. The end of Ramazan follows Turkish custom, with Seker Bayram (Aid ul-Fitr), a three-day holiday where greetings cards are sent and guests received and served with coffee and baklava. The next feast after 70 days is Kurban Bayram (Aid ul-Adha) when it is customary to kill a sheep for a celebratory meal. For Romanian Muslims, observation of the fast is an individual decision as there are no public adjustments to daily life.

Pilgrimage takes place annually in Dhul Hija, the twelfth month of the Islamic year. Muslims who can afford to do so are obliged to undertake the pilgrimage to Mecca and Medina at least once in a lifetime. Traditionally this involved a long, hard journey overland. Pilgrimage was a problem because of the uncertainty of foreign travel. From time to time, however, a delegation was allowed to visit Mecca led by a member of the Muftiate staff who had studied in Saudi Arabia and knew Arabic well.

Apart from orthodox religious beliefs and obligations, Muslims share the civic duties of all Romanian citizens and, to a certain extent, their life-cycle rituals. In the cities at least, all Romanian Muslims are multilingual, speaking Tatar, Turkish and Romanian and are therefore able to manage in a dominantly Romanian society. In general, they blend physically, having the dark hair and eyes of Romanians, although some have a distinctive Tatar appearance with high cheekbones and prominent noses. Most of them wear Western dress, although some women still wear colourful, voluminous ankle-length trousers with matching blouses and have their long hair plaited and tied up under headscarves. Women have not adapted to the fundamentalist, all-concealing 'hejab' dress and face veil. Men's dress includes the traditional styles as worn by Imams in the mosque of a 'cuibe': a long black coat and a turban usually of white or yellow cloth embroidered with Arabic inscriptions and wrapped around a fez; a green cloth indicates that the wearer has been on pilgrimage to Mecca. A few tailors and hatmakers in Constanta specialise in the making of such garments.

Degrees of belief among Romanian Muslims are variable. Older people try to maintain all aspects of belief and practice but younger people are more flexible, reserving Muslim traditions and customs for family occasions. This dualism also runs through education. Romanian Muslims enter the state system of primary and secondary schools and university. Any religious instruction is through the home or controlled by the state-approved Muftiate. Koran study is a revealing example of this situation. Muslims accept that the Koran is the word of Allah transmitted in Arabic through the prophet Muhammad which guides all aspects of a believer's life. It is therefore important that children are taught to read the Koran and to learn it by heart. This was a problem in Romania as copies of the Koran were limited. Such as were available were not printed in Romania but imported from Turkey to the Constanta Muftiate which controlled their distribution. The position of the traditional Koran school, 'mektub', was also problematic as the Romanian state had firmly discouraged such religious teaching. A solution was found by teaching Muslim children Arabic and the Koran twice a week in the main Constanta mosque, after they had finished state school for the day. A comparable situation existed concerning specialised university-level education. Traditionally, the Muslim institution of the 'medresse' – which funded instruction and accommodation – trained candidates for religious and legal posts. The closure of the last 'medresse' in Medjidia in 1964 meant that there was no provision for specialised Muslim education in Romania. Various reasons were given for this: applications had dwindled to such an extent that the medresses had become redundant; the ratio of about 120 Imams to service the 50 or so Muslim towns and villages was thought to be sufficient; many of these Imams were also young and some had been trained in Saudi Arabia, so there was no immediate staff shortage.[5]

Muslim life-cycle rituals were and are easily accommodated within a secular framework and to a certain extent are influenced by Romanian custom. The circumcision of boys between the ages of 6 and 8 is unique to Muslims and is treated as a family party. As in Turkey, the boy is dressed in a smart white military suit, complete with peaked cap, bandolier and cape in which he parades among admiring relatives and friends. After his circumcision, he is dressed in a long white shirt and put into an elaborately decorated bed where he continues to hold court and receive gifts of money to the background accompaniment of musicians playing the traditional tambourine, 'def', and clarinet, 'zurna'. The next significant step is marriage where there is more opportunity for a mingling of Muslim and contemporary Romanian practice. The choice of marriage partner is generally Muslim and, while a couple have freedom of choice, they usually consult parents and other relatives. A marriage contract is prepared, signed and witnessed, which in Muslim terms means that the couple are now married and can live together, though in practice they delay this. Their status is legalised when the couple register their marriage at the local municipality which is obligatory for all Romanian citizens. While lavish and noisy wedding parties are enjoyed by all Romanians, those of Muslims are celebrated three months after the signing of the contract, a period equivalent to an engagement. The bride wears a white European-style wedding dress to which the guests pin gifts of money. Both families provide generous hospitality, including musical entertainment and Turkish-influenced food such as 'yesil fasulya', green beans, a selection of 'pilav', rice dishes, and 'yaprak dolmasi', stuffed vine leaves. The last rites of all follow Muslim tradition only. Immediately after death, the body is washed, wrapped in a white shroud, and taken quickly to the cemetery for burial by sunset. The funeral is followed by a series of Koran readings at the graveside, and then at intervals of 3, 7, 37, 40, 52 and 100 days after death and thereafter annually.[6] A simple monument of two stones inscribed in Arabic or Latin script is placed, one each at the head and foot of the grave.

Intermingled with the formal structures of Islam, and life-cycle rituals adapted to Romanian state requirements, are occasional and seasonal celebrations. One of the most popular is a reading of passages from the 'Mevlut', a long narrative poem relating of the life of the Prophet Muhammad. 'Mevlut' readings are given at home to invited guests who are served coffee and sweets, to celebrate the birth of a child, success in examinations, the purchase of a new home, and the like. These 'Mevlut' practices are similar to those of Turkey. Specific Tartar festivals celebrating the Spring sowing and Autumn harvest are found only in villages. Families traditionally go to the fields and spend the day enjoying a picnic meal which includes a festive dish of meat cooked in crisp layers of paper-thin pastry and listening to Koran recitals.

Survival of the Muslim community in modern Romania is the result of a cautious balance between the potentially conflicting requirements of Islam and the state. Various factors have contributed to this situation. The community is small and contained in one area, the Dobruja. Its members speak Romanian in addition to their Turkish and Tatar dialects, participate in the state educational system and generally blend with the majority Romanian population in appearance and dress. In the towns they do not live in segregated areas and mix socially with their Romanian neighbours, attending, for example, each other's weddings,

Christmas and Bayram parties. The relative simplicity of Islam in terms of ritual and clerical structure contributes towards a discreet practice of religion. Romania's Muslims, however, are not isolated as they are members of a major faith which transcends national boundaries. This has been of advantage to the Romanian state which has cultivated relationships with Arab socialist states – Algeria, Iraq, Syria – which has involved contracts and agreements at all levels from receiving delegates, enrolment of students in Romanian universities, and so on. Here, the Muftiate in Constanta has been able to offer specialist skills through members of its staff who have been trained in Arabic and Islamic studies. They are able to receive, entertain and interpret for Arab diplomatic and trade delegates, and organise induction courses for Arab students. Provision of these services has created some opportunities for the Muslim community. Groups of Muslims were at intervals permitted to travel on pilgrimage to Mecca, escorted by Muftiate representatives, who in turn were able themselves to attend meetings of the Islamic Conference which funded their expenses. In effect, a mutually advantageous and dynamic network evolved between state and community. The position of Romania's Muslims after the 1989 revolution stimulates further questions of survival. As Romania and neighbouring territories experience a relative liberalisation of state control, have Muslims responded by further emigration? If so, what are their preferred destinations? Turkey is an obvious choice as many already have family contacts through the communities which emigrated in the early twentieth century. If they choose to remain, do they follow a route of increased secularisation and assimilation into the Romanian community or do they attempt to define a more public Muslim identity? There is plenty of opportunity for research into this intriguing minority.

Notes

1 My research on Romania's Muslims is the result of a long-established interest in the Muslim communities in the Balkans which in turn developed out of my studies of the Ottoman Turks and their extraordinary and steady penetration of Central and South-Eastern Europe. I became curious about both the nature of Ottoman rule in these areas and the impact of Islam on local cultures. I have visited Romania frequently since 1975 and had the opportunity of doing fieldwork in 1980 and 1986 on the carpet industry which combines both indigenous weaving traditions with imported Turkish techniques. Within this fieldwork, I was able to pursue my interest in the Muslim community and was fortunate enough to interview extensively in 1986 and on later occasions one of the Imams of the Constanta Muftiate. Apart from the historical background, the information presented here is based on my fieldwork notes.

2   The pattern of immigration is complex and often confusing and requires further critical study. Bibliography is sparse and elusive. A useful account which I consulted is Frederick de Jong, 'The Turks and Tatars in Romania', *Turcica Revue d'études Turques*, Tome XVIII, 1986, pp. 165–89.

3   This was the ethnic background of my informant, whose family had been important in the Dobruja. His father had been Mayor of Medjidia, while his grandfathers had been Muslim jurists.

4   The Directorate of Historical Monuments restored the mosque from 1957 to 1958. During a visit to Romania in 1998, I was informed that it was closed for repair and was enclosed in scaffolding.

5   My informant was young, in his mid-30s and well-educated. He had finished his Islamic studies in Saudi Arabia and spoke excellent English in addition to Romanian and his working and local Arabic, Tatar and Turkish languages.

6   The significance of numbers in post-funeral ritual seems to reflect arbitrary choice. They do, however, represent logical units of time with increasing intervals between them as the intensity of mourning decreases. According to my informant, the 37th and 52nd days are based on the numerology of the Tibetan Book of the Dead. This is a curious statement which as yet I have not been able to check against the funerary practices of other Muslims. The origin may be found in Central Asia where Muslim and Buddhist communities have co-existed. As with so many aspects of Romanian Islam, the problem requires further investigation.

11
The Bektashis in the Balkans

John Norton

Introduction

The Bektashis became the most influential of the dervish orders established in the Balkans under Ottoman rule. Yet, although they originated in Turkey, they were frequently persecuted by the Ottoman establishment. This chapter will endeavour to outline the origins, development, beliefs and practices of this Sufi order in Turkey and in the Balkans and explain why it was often vilified and victimised.

Inevitably the generalisations used to describe any such association of millions of human beings that has lasted for centuries and spread across continents will distort the views of some of its individual members. This is particularly true of the Bektashi order, which has been receptive to ideas from so many quarters and is renowned for its tolerance. Thus the teachings outlined below will not necessarily be those of Bektashis everywhere at all times. Nevertheless, they are intended to convey a reasonable general impression that most Bektashis could accept.

The Sufi setting

The Bektashis are but one – albeit a very distinctive one – of the many bands of believers seeking to achieve the goal of most Sufis: mystic union with God. It may therefore be helpful first to note the general features of Sufi orders before turning to the particular characteristics of the Bektashi order.

The original Sufis were individuals whose craving for spiritual satisfaction led them beyond the austere demands and prohibitions of the Islamic faith that had been but recently proclaimed by the Prophet. Finding inspiration in interpretations revealing the 'inner meaning' behind the literal sense of sacred texts, these early Sufis formed schools of Sufi doctrine. These, as well as different schools teaching other branches of Muslim thought, had sprung up within the first two centuries of Islam.[1] In course of time the various methods or ways that these schools found to attain their goals developed into separate orders of dervishes each being known in Arabic as a *tariqa* (way), which in Turkish becomes *tarikat*.

As well as developing its own distinctive approach, every *tarikat* displayed features that were common to all of them.[2] These common features included:

- controversial interpretations (*ta'wil*) of Koranic texts and traditions of the Prophet (*hadith*) to justify their conduct and beliefs. For Sufis, because they sought to lose themselves in divine love, the Koranic text that appealed above all others was 5:59, 'A people whom he loveth and who love Him.'[3]
- a genealogical chain (*isnad*) linking their founder to the Prophet and other authorities.
- an organisational framework with a leader who derived his authority from the founder in either genealogical or spiritual succession.
- a rank structure with various grades below the leader.
- the appointment of spiritual guides (Turkish singular: *mürşit*) to teach those who opt to follow the way (*muhip*).
- initiation rites.
- the total and blind obedience owed by the *muhip* to the *mürşit*.
- various stages of enlightenment along the way to ecstatic union with God.
- dervish lodges (known by various names, in Turkish commonly *tekke* or *dergâh* or *zaviye*), often sited near the tomb of the order's founder or other venerated saints.
- worship and ceremonies conducted in these dervish lodges.
- headgear and other forms of dress peculiar to the order.
- lay members who would meet with the full-time members for worship at the lodges and then return to their normal occupations outside.
- in varying degrees, the suspicion, hostility or outright condemnation of the orthodox Islamic establishment, which held that Koranic law legislated only for an external tribunal and claimed that the mystics were guilty of heterodoxy. (Sufis acknowledged an inner tribunal by which to judge themselves and others and to purify their consciences. They held that intention and practical example were more important than the strict letter of the law.)

Origins of the Bektashi order

Before considering how those common features of Sufi orders are reflected in the Bektashi order, we should briefly note how the order came into being.

The historical facts concerning the life and deeds of Haji Bektash, whom the Bektashis revere as their founder and patron saint, are clouded in obscurity but there is reasonable documentary evidence of his existence in the thirteenth-fourteenth century AD. He is thought by some historians to have been a disciple of Baba İshak,[4] whose rebellion in 1239 threatened to overthrow the Seljuk Sultan, Giyasuddin Keyhüsrev II. (If this was the first Bektashi link with insurrection, it was certainly not the last. In subsequent centuries Bektashis gained a reputation as stalwart supporters of rebellions against oppressive authorities.) It is not known for certain that Haji Bektash personally founded the order or whether his name later came to be associated with it. However, as with most religions, legends accepted by believers are much more influential than mundane facts. According to the legend related by most Bektashis today, Haji Bektash was born in AD 1248 in Khorasan (the province which is now a north-eastern province in Iran). Khorasan was also the home of Ahmad Yasawi (known in Turkish as Ahmet Yesevi) (d. 1166), to whose charismatic inspiration a number of Sufi orders can be traced. Despite the anachronism, the popular legend asserts it was at Ahmet Yesevi's personal behest that Haji Bektash set off in 1281 to spread the faith in central Anatolia. Around the same time, other dervishes influenced by Yesevi teachings also moved into Anatolia along with many ordinary Türkmen who were migrating westwards before advancing Mongols. The Türkmen proved valuable warriors for Seljuk sultans and their Ottoman successors, though their fierce nomadic ways frequently put them at odds with their urbanised rulers who had adopted Sunni Islam.

Haji Bektash established his centre of operations between Nevşehir and Kirşehir in a tiny village called Sulucakarahöyük, which has today become the small town that bears his name: Hacibektaş. Bektashi legend claims he wrought numerous miracles, healed the sick, and won many converts. Further legend, contrary to historical fact, links him to the founding of the Janissaries in 1295. He is said to have died in 1337. Some of these dates are derived from a summary of his life that in Hurufi tradition endows key letters with numerical values and symbolic significance.[5] A different account gives his date of birth as 1207, his move to Anatolia around 1230, and his death, at the age of 63, in 1270.[6] Numerous variations give yet other dates, so that there is a difference of up to 40 years between the birth dates given in various sources. Similarly, there is a difference of up to 68 years in the dates of his death.[7]

It appears that Bektashis absorbed a number of similar groups of dervishes variously known as *abdals, haydaris* and *kalenderis,* most of whom were more or less opposed to authority.[8]

After its foundation, at whatever date, the development of the Bektashi order falls into two distinct periods: before 1500 – formation; after 1500 – regulation and organisation.[9]

Before 1500 both the order and its teachings developed in a random fashion. Türkmen tribesmen who found Bektashism attractive had brought with them an amalgam of shamanistic, Buddhist and Manichaean as well as Islamic beliefs and practices. In Anatolia, and later in the Balkans, they encountered Neoplatonist, Christian and Nestorian influences too. Bektashism adopted elements of all these religions and was also receptive to Kalenderi and Hurufi ideas. After exposure to Shia teaching, Bektashis came to revere above all else Ali, the fourth Caliph of Islam, whom Shiites held to be the first rightful successor to the Prophet. And, with the adoration of Ali, Bektashis also adopted the associated practice known as *teberra,* execration of Ali's enemies, particularly the first three caliphs. Syncretism made Bektashism acceptable to many non-Muslims who would have found the austere teachings of Sunni orthodoxy unattractive. But their Shia sympathies further distanced them from the Ottoman establishment, which was Sunni. Moreover, Bektashi readiness to incorporate such a wide range of beliefs and practices made it difficult to define precisely what the order stood for. This vagueness of belief was paralleled by lack of uniformity in their organisation.

Around the year 1500 Balim Sultan (1473–1516), the baba of an important tekke in Dimetoka (Didimotikhon), some 22 miles south of Edirne (Adrianople), was appointed leader of the order and took up his post at its headquarters, the tekke in Hacibektaş. He took firm control and introduced a large degree of uniformity. For this reason he is revered as the Bektashis' second founding saint. He formalised a belief system that propounded a concept of a trinity comprising God (who is referred to in this connection as *Hak* – Reality), Muhammad and Ali. Although God was described as the one Reality, both Muhammad and Ali were regarded as special manifestations of this Reality.[10] The Twelve Imams revered by Shiites were also elevated to a place of great importance in Bektashi doctrine, with Dja'far Al-Sadik (in Turkish Cafer Sadik) (699–756), the sixth Imam, being claimed as the one from whom they derived their authority.[11] (The perceived parallel to the Christian Trinity and the Twelve Disciples has led many commentators to presume that these features were taken from Christianity and simply given a Bektashi veneer. Bektashis deny this.)[12]

Although Balim Sultan set the order on strong foundations, it was from his time that a split developed in its ranks between the group known as *Yol Evlâdi* (Sons of the Way), who held that Haji Bektash never married and had no children so leadership of the order should go to those deemed most worthy of the honour, and the rival group known as *Bel Evlâdi* (physical descendants), who claimed he did marry and that leadership rightly belonged to the person with the strongest genealogical claim. It was the *Yol Evlâdi* who attracted converts in the Balkans.

Features of the Bektashi order

We can now examine how the general features of Sufi orders manifest themselves in the Bektashi order.

- *The use of Koranic authority.* Bektashis appeal rather less to Koranic authority than do most other orders. The verses and traditions they select for special emphasis relate chiefly to the exalted position of humans above all other creation, and to the themes of love, the importance of Ali, or the 'unity of existence', a very important Bektashi theme.[13] Appropriate Koranic verses are read at weddings and funerals, but otherwise Bektashis are happier to listen to the verses of their poets conveying their own responses to the sacred texts.
- *Lineage.* Bektashi genealogies trace Haji Bektash's line back through, among others, Dja'far Al-Sadik (Cafer Sadik), the sixth of the Twelve Imams revered by Shiites, Hüseyin, the grandson of the Prophet, and Fatima, the Prophet's daughter and wife of Ali, and thus to the Prophet Muhammad.
- *Organisation.* As already noted, Balim Sultan purged and reorganised the order. Although Bektashism had spread over a wide area and gained many adherents before 1500, its tendency to syncretism had led to a lack of theological and organisational uniformity. Balim Sultan changed that by exerting strong control from the headquarters of the order, the tekke at Hacibektaş.
- *Rank structure.* Balim Sultan instituted a rank structure that has persisted. The head of the order resided at the central tekke at Hacibektaş, and directed its affairs from there. After the split between the

Yol Evlâdi and the *Bel Evlâdi* developed in the sixteenth century, each group had its own leader resident there. The head of the *Yol Evlâdi* was called the *Dedebaba*, and the *Bel Evlâdi's* leader was known as the *Çelebi*. The heads of the other tekkes held the rank of *baba* and between them and the supreme head at the Hacibektaş tekke four *halife* were appointed to oversee the babas within their geographical area, in the same way that a Christian bishop oversees the priests in his diocese. Fully initiated members were called *dervish*, those at the lower level were termed *muhip*, and aspirant members yet to be initiated were called *aşik*. An *aşik* could participate in some but not all of the activities at the tekke.

- *Spiritual guides.* A *mürşit* in the Bektashi order carried immense prestige and responsibility. It was his task to judge the personality and capability of *muhips*, instruct them in the teachings of the order and at appropriate stages reveal to them the order's secrets. In language strongly reminiscent of Chapter 3 of St John's Gospel, the person many Bektashis currently revere as head of the order writes thus of the vital relationship between a Bektashi *mürşit* and *muhip*: [A Bektashi] 'is born first via his mother and then born again into a new world via his *mürşit*. He who is not born twice cannot enter the kingdom of the heart and he sees mankind as distinct from God and God as distinct from mankind'.[14]

- *Initiation rites.* The *aşik* who has received instruction and is deemed worthy by his *mürşit* is initiated into the Bektashi order at a special service known as the *ikrar ayini*.

- *Obedience to the mürşit.* The Bektashi is taught that he owes a greater allegiance to his *mürşit* than to his own parents, and what the Prophet is to the community of Muslims so the *mürşit* is to the *muhip*.[15] Failure to obey the *mürşit* is a failure to keep the oath sworn at initiation. Obedience must be total and unquestioning. (Clearly, in such a relationship an unworthy and unscrupulous *mürşit* will be tempted to exploit his power for personal advantage. From time to time such abuse occurred, but it was condemned by Bektashis and outsiders alike.) [16]

- *Stages of enlightenment.* Bektashis teach that there are four stages of enlightenment. These are often called the four gateways and are known as Şeriat, Tarikat, Marifet and Hakikat. They marked stages of spiritual development from initiation right through to the immediate experience of Reality. The first of these gateways, Şeriat (Sharia) is, as its name describes, the formal law of Islam. But since most Bektashis are notorious for flouting these laws, it may seem strange to the outsider that this should be a requirement. The approach many Bektashis take is that the external facts about the faith were taught by Muhammad and Muslims should be aware of them, but it is through Ali they can gain insight of the inner meaning these facts signify. Indeed, there are four layers of meaning: *external* for the common people; *subtleties* for the gnostics; *secrets* for the saints; and inner truth or *real essence* for prophets.[17] Progress through the four gateways leads the Bektashi from the formal requirements of Islam through the mysteries of the dervish order (tarikat), into mystic knowledge of God (marifet) and finally to immediate experience of the essence of Reality (Hakikat).

- *Tekkes.* In their tekkes the Bektashis conducted their ceremonies, managed the business of the order and entertained guests. It was claimed that Bektashi tekkes were once so numerous that a traveller crossing the Ottoman Empire would always find one within a day's journey.[18] Often they were sited in imposing locations. Within the tekke there would be a large room, called the *meydan* or the *kirklar meydani*, where the ceremonies were conducted, a cookhouse, a bake-house, pantries, guest rooms, stables, and at some tekkes there would also be the tomb (*türbe*) of a saint. Today tekkes are to be found as far away as North America where Bektashis have settled, and Bektashis in America have helped Albanian Bektashis rebuild and reopen tekkes since the fall of communism.

- *Worship and ceremonies.* Special services and ceremonies were held when initiates were received and when members of the order pro-gressed to a higher grade. These ceremonies tended to be long and complex and involved set prayers or special verses (called *gülbenk* and *tercüman*) some of which were recited by everyone but others were reserved for persons with a particular function. Members of the order could attend the ceremony welcoming a new person up to and including their own rank, but such gatherings were not open to those to whom a *mürşit* had not yet revealed the secrets relating to that level.[19] There used to be a special ceremony for Bektashis who wished to avow celibacy. On the threshold of the tekke they would have a hole bored in an ear and thereafter they would wear a special earring.[20]

- The act of prayer at Bektashi worship differs from the standard Muslim *namaz*. Bektashis are taught that inner purity and sincerity are more important than outward ablution and ritual. As it is con-sidered important that everyone should understand, most of the prayers and worship are conducted not in Arabic but in the language of the congregation. Among the various Bektashi ceremonies is an annual one for the confession of sins, righting of wrongs and granting absolution.[21] Members who are blameworthy may be awarded a variety of punishments, the most serious of which is excommunication. The isolation resulting from excommunication has been known to drive victims to suicide. But it is important to note that Bektashis hold out neither the blandishments of a future Heaven nor the torments of a Hell in the Hereafter to promote good behaviour in this world. Death may be regarded as simply a return to God the Reality and is often referred to as 'migration'.

- Other occasions for special ceremonies include *Nevroz*, the Persian New Year, which is considered to be the birthday of Ali. And in the first ten days of the Islamic month of *Muharrem* they hold mourning nights to commemorate the suffering and death of Hüseyin (625–82), the second son of Ali. On the evening of the tenth day the ceremony of 'Ashura' (in Turkish: Aşure Merasimi) is held. A special dish, called aşure, made of wheat, hazel-nuts, raisins, almonds and other ingredients is cooked in a large cauldron, stirred in order of rank by the Bektashis present, and distributed to all around. (Cauldrons acquired symbolic significance among Bektashis and Janissaries. When Janissaries overturned their cauldrons, rulers trembled because that action denoted revolt.)

- *Distinctive dress.* Bektashis have several items of distinctive dress, all endowed with symbolic significance. The standard patterns of their headdress usually have twelve or four pleats or ridges indicating the Twelve Imams and the Four Gateways respectively. A twelve-fluted stone of onyx or crystal hung round the neck or fastened to a cummerbund is known as the *teslim taşi* and symbolises submission to God.

- *Lay members.* In addition to full-time dervishes who would mostly live in tekkes, many Bektashis were lay members who followed ordinary careers but took part in ceremonies in tekkes, confessed their sins annually and in general endeavoured to follow Bektashi precepts in their everyday lives.

Incurring the hostility of the orthodox establishment

In addition to the criticism directed at all Sufis on account of their deviations from Sunni orthodoxy, Bektashis drew special wrath upon themselves and were frequently denounced as heretics because of their seeming disdain for the 'Five Pillars of Islam' – the five essential duties that Sunnis believe all Muslims must accept. The first of these is the *testimony of faith*, 'There is no God but God, and Muhammad is God's messenger'. To this statement Bektashis add further words declaring that Ali is the companion of God and the Guardian of the Prophet.[22] The next pillar is *namaz*, the ritual act of prayer to be performed five times daily. Most Bektashis do not feel bound by this obligation, which they see as outward show, and they emphasise that they strive for inner purity. In any case, they do not go to mosques to perform *namaz*. If they perform it all they do so in the privacy of their own homes.[23] Nor do they feel the need to face Mecca when they pray. Their tekkes have no *qibla* to indicate the direction of Mecca. As for *zakat*, almsgiving, Bektashis say the obligation they have to support those in need – an obligation the babas are duty-bound to see observed – renders superfluous the requirement to offer a fixed percentage of their income. With regard to *oruç*, fasting, Bektashis were in constant trouble for failing to observe this during the month of Ramadan. As already noted, they fast instead for a time in the month of Muharrem.[24] As for *haj*, the pilgrimage to Mecca that every Muslim who can afford it is enjoined to make at least once in a lifetime, Bektashis go instead to Hacibektaş,[25] where they may follow certain customs, some of which are reminiscent of Meccan pilgrimage rituals and others that are of pre-Islamic origin and feature trees and rocks. But, as well as being condemned by the religious authorities for their laxity, Bektashis were usually subject to the suspicion and animosity of the *political* authorities too because they traditionally sided with the underdogs and frequently joined rebellions.

Further distinctive features of the Bektashi order

To give a sufficiently comprehensive picture of the Bektashi order the details so far presented with reference to how Bektashis fit into the general Sufi framework must be supplemented by some comments on certain aspects of beliefs and practices peculiar to the Bektashi order.

- *Bektashis and Shiism*. Bektashis are often described as Shiites. This may give a misleading impression. Although, as we have seen, they share with all Shiites extreme reverence for the Caliph Ali, and stress their allegiance to the *ehl-i beyt* (the Prophet's family), Bektashis are very different from the Shiites in Iran, who are commonly thought of as being representative of that branch of Islam. Iranian Shiites worship in mosques. Bektashis do not, and they actually resent having mosques foisted upon them by interfering authorities. Iranian Shiites under the influence of Khomeini abhor alcohol and seek to ban it. Bektashis delight in drinking alcohol. Drinking has an important place in many of their ceremonies; they also happily imbibe alcohol on social occasions. Iranian Shiites strive to veil and seclude women. Bektashis do not; they proudly claim to treat women as equals; they stress a saying attributed to Haji Bektash: 'Educate women! A people that does not educate its women cannot progress!' Although not all Bektashi women have yet achieved the degree of equality to which the women's liberation movement in the West might aspire, they mix freely with men in everyday life and in the *sema* (the traditional Bektashi dances full of religious symbolism) they dance together with men. Moreover, Khomeini's followers search the *Sharia* to back their imposition of a narrow and restrictive code of conduct. Bektashis condemn this approach in the same terms that Jesus condemned the scribes and Pharisees for preferring the letter to the spirit of the law. The Bektashi code of conduct is simply expressed in Turkish: *Eline, Diline, Beline Sahip Ol!* – a straightforward command to exercise self-control to avoid harming anyone by word or deed. This, they argue, is an adequate guide to conduct. Thus Bektashis are by no means typical Shiites.

- *Bektashis and humanism.* Indeed, Bektashis are so far removed from standard Shiism that they are quite often described as humanists. But this too is a misleading appellation, though many of them are happy to accept it. Like humanists, they see love of fellow humans as the highest ethic and they attach the highest value to humanity, so they may feel they are heading in the same direction as humanists. Indeed, the great Albanian Bektashi poet, Naim Frashëri (1848–1900) said that 'God is mankind'.[26] But the additional baggage of Bektashi beliefs and the effect upon interpersonal relationships of the order's hierarchical structure (particularly that between *mürşit* and *muhip*) make them far from typical humanists.

- *Bektashis and pantheism.* Often, Bektashis are called pantheists. To the extent that this indicates their view that there is no radical distinction between the Creator and his creation and that everything exists within a single reality, this is a fair, though by no means comprehensive description.

- *Bektashis and metempsychosis.* It is frequently said that Bektashis believe in the transmigration of souls. This is not, however, an important theme in their teaching and in practice it would appear to be something that not all of them accept. Bedri Noyan, a leading Bektashi, says on this subject that while all creatures may exhibit some aspect of God, it is only in human beings that all the qualities of God are made manifest. Before attaining that level, a human will have passed through all other lower forms of life. The human soul is like a caged bird, striving to be free to be united with Reality. For one with a mature soul advancing to Reality there is no turning back, no more suffering the pain of separation. The believer never dies; he passes from this transitory world to an eternal world. For Bektashis death is not separation, it is union; it is birth into a new world. It is the illuminated road that takes the person to his beloved. 'The Bektashi belief beautifies death and thus offers humans the greatest consolation.'[27] He does, however, state that Bektashis are above all people of the day that is being lived. Their concern is to spend that day well, beneficially and beautifully. Nevertheless, some believe that when a life has not been lived well, metempsychosis may take an unhappy turn. It was said to be a custom of some Balkan Bektashis to avoid treading on insects lest they should have been Bektashis in a previous existence.[28] Good people were more likely to become birds. Legendary leaders of the Bektashis did not have to await death before assuming different form. They could do so at will. Haji Bektash, for example, is said to have approached Sulucakarahöyük in the guise of a dove. Such abilities were the stock-in-trade of shamans. As Professor Norris points out, these beliefs were prevalent among pre-Islamic Turks.[29]

- *Mysteriousness and numerology.* Bektashis have always been prone to clothe their mystic teachings in mysteriousness. Those Bektashis who were influenced by Hurufism made great use of numbers and the symbolic significance attached to letters of the Arabic alphabet to illustrate their message. Often this practice would be taken a stage further to circumvent the ban on drawings representing living creatures: letters would be combined to form the shape of human faces and bodies or to resemble animals. Numbers of special importance include: 3 (the trinity of Allah, Muhammad and Ali), 4 (the 'Four Gateways'), 5 (members of the family of the house of the Prophet (in Turkish *ehl-i beyt*): Muhammad, Ali, Fatima, Hasan and Huseyin – frequently represented in the charm commonly known as the 'Hand of Fatima'), 12 (the Twelve Imams; and, to add to its importance, it is also the sum of 3, 4 and 5!), 14 ('The Fourteen Pure Innocents', all children of the Twelve Imams, including some killed at the battle of Kerbela; the existence of fourteen joints on the five fingers of one hand is said to demonstrate the significance of this number), 40 (the Forty with whom Muhammad conversed during the *miraj* – his ascent to heaven), 72 ('the seventy-two religious communities' – from a tradition of the Prophet that says, 'My people will be divided into seventy-two sects; all of them destined for hell except one, and these are the true believers'). At the *tarikat* stage of enlightenment the one approved sect is that of Dja'far Al-Sadik (Cafer Sadik). But at the *Hakikat* stage all differences can be overlooked.[30]

- *Secrecy.* Secrecy has always been a feature of Bektashism. It is born partly of persecution and partly of mysteriousness. The secret signs by which Bektashis, like Freemasons, could recognise one another without revealing their allegiance to outsiders exemplified this characteristic. It is apparent too in the frequent references to the 'Bektashi secret', about which so much has been written. It used to be said that Bektashis were on pain of death not to reveal their secret. That claim gave rise to much wild speculation about what the secret really was. Varying answers have been given but it is likely that deliberate dissimulation, *takiye*, makes the reliability of some of these questionable. It is often claimed that the secret really is that God as a separate entity does not exist – a secret too dangerous to reveal to hostile Sunnis.
- *Bektashis and freemasonry.* Bektashis themselves sometimes mention certain similarities between their order and freemasonry. Some individuals were members of both. Nevertheless, it is incorrect to assume any formal link between the Bektashism and freemasonry.[31]
- *Bektashi humour.* A strong feature of Bektahism is humour. Very often it is the humour of the oppressed; laughter in the face of adversity. It serves to reinforce Bektashi social solidarity. Over the centuries, Bektashis have told countless jokes about punishments meted out to them for failure to comply with the rules and customs of orthodox Islam. Some of their jokes adopt a reproachful attitude towards the Almighty and can appear blasphemous. They certainly reflect a relationship to God that is far removed from fear and dread. Love for God and for fellow man and hatred of hypocrisy are frequent messages that Bektashis seek to convey through their humour.
- *The written word.* We have already noted that Bektashis do not spend long hours immersed in the Koran. Their beliefs have traditionally been transmitted orally, with the *muhip* receiving instruction direct from the *mürşit* and with poets and minstrels reciting and singing the essence of their message to all who are willing to listen. Among the most quoted teachings attributed to Haji Bektash are: 'Whatever it is you seek, seek it within yourself'; 'Do not impose on others anything that you would find too hard for yourself'; 'Be the balm for wounded

hearts'; 'The road that does not pass through knowledge leads to darkness'; 'Blessed is he who shines light into the dark places of thought'; 'Use your power not against the underprivileged but against the cruel'. The written sources for their traditions are limited and are mostly collections of their legends rather than closely argued expositions of their faith. The *Vilayetname* (Book of Saintship) and the *Makalât* (Sayings of Haji Bektash) are the most important of these collections. *Hüsniye* is another centuries-old work that Bektashis treasure. It recounts the legend of a sinless slave-girl whose wisdom exceeded that of the *ulema*. In recent years there has been a spate of books in Turkish about Bektashism, many of them by Bektashis, and some of them offering new interpretations and political slants. While these may provide material for lively debate, extracts from Naim Frashëri's Bektashi Pages,[32] first published in 1896, provide a useful example for our purposes, since they emphasise several of the important aspects of Bektashism, particularly in its Balkan manifestation:

For their first founder they hold Jafer Sadik and for their patron saint Haji Bektash Veli, who is descended from the same family.

For the Bektashi the Universe is God. But in this world man is the representative of God. The True God, with the angels and Paradise and all that is good, are found in the virtues of man. In his vices are found the Devil and all evil.

As the man, so is the woman, one in kind and not separated.

The woman does not veil or cover her face save only with the veil of modesty.

The Bektashi have for the book of their faith the Universe, and especially mankind... Faith is in the heart, it is not in the written book.

Not only among themselves but also with all men the Bektashi are spiritual brothers. They love as themselves their neighbours, both Mussulman and Christian, and they conduct themselves blamelessly towards all humanity. But more than all they love their country and their countrymen, because this is the fairest of all virtues.

The Poor [brother] if he is married before he takes the habit, may remain in wedlock after his election, abiding in his family and in his house. But when he takes the vow never to marry, he obtains a new Permission but he cannot take back his word. The unmarried Poor [brothers] live in a house which is called *tekke* or *dargah*.

But sometimes there are many Grandfathers: then they choose from among them and make him Great Grandfather.

Guides must all be men perfect in all things pertaining to the Way. Before all things love is an approach and an interpretation of the Way.

For a fast they have the mourning they keep for the Passion of Kerbela, the first ten days of the month which is called Muharrem. In these days some do not drink water, but this is excessive, since on the evening of the ninth day the warfare ceased, and it was not till the tenth after midday that the Imam Husain fell with his men, and then only they were without water. For this reason the fast is kept for ten days, but abstention from water is practised only from the evening of the ninth till the afternoon of the tenth. But let whoso will abstain from water while he fasts. The Way of the Bektashi holds all men, yea, all men, friends, and looks on them as one soul and one body.

The true Bektashi respects a man of whatever religion he may be, they hold him their brother and their beloved, they never look on him as a stranger. They reject no religion but respect all.

Janissary connections

It was in large part due to their military connections that the Bektashis achieved an importance in the Ottoman Empire far greater than might have been expected of a dervish order with unorthodox views. We have noted that Türkmen *gazis* who spearheaded the expansion of the empire in its early days felt an affinity with the Bektashis. Further political importance was gained by the Bektashi connection with the Janissaries. This connection began before 1500 and continued until 1826. It contributed greatly to the order's success in establishing itself in the Balkans.

The Janissaries were the personal troops of the Ottoman sultan. Originally formed from prisoners of war retained by Murat I (1326–89, ruled 1360–89) as his *pençik* (his one-fifth share of booty captured from the enemy),[33] they soon came to be recruited through the *devshirme* system (conscription of Christian boys from the Balkans in levies occurring at intervals of between one and seven years).[34] Fit young men and boys were taken to Istanbul or Bursa, sorted according to aptitude, converted to Islam, and those destined for military service were given appropriate training before enrolment as Janissaries. Writing in the seventeenth century, Ottaviano Bon, a Venetian envoy in Istanbul, described the process as follows:

They are taken from such families as are supposed to be of the best spirit, and most warlike disposition; nor may they, when they are gathered, exceed twelve or fourteen years of age at the most, lest they should be unfit for a new course of life, and too well settled in Christianity to become good Turks. The *Capooches*, having finished their circuit, and gathered their whole complement, bring them forthwith to Constantinople...Then they are circumcised, and made Turks, and set to learn the Turkish tongue... generally all of them are taught to wrestle, to leap, to run, to throw the iron weight, to shoot the bow, to discharge a piece, and, to conclude, all such exercises as are befitting a Turkish soldier.[35]

So long as men were Janissaries they were the personal slaves of the sultan and under his direct command. Originally, they were not allowed to marry before retirement. Despite the seeming harshness of this recruitment system, once it was well established many families were eager to have their sons accepted because a successful career could lead to the high offices of state or to positions where they could bestow favours on their home villages.[36] The Janissaries became the élite infantry of the Ottoman army. Their numbers varied between 5000 and 54 000.[37] In time of peace, they represented the sultan's central authority in the provinces. In time of war, with their colourful uniforms, distinctive march, military bands, and fearsome weapons, they achieved a formidable reputation and their approach was feared, particularly in Europe, where twice (in 1529 and 1683) they led Ottoman armies to the gates of Vienna. But decline set in. The last full *devshirme* levy in Bulgarian lands was in 1685.[38] Although in their early days, Janissaries had been a useful tool for the sultan, enabling him to withstand rival elements that were competing for power, they later degenerated and fought for their own interests rather than those of the sultan or of the state. Indeed, they also gained the power to make and break sultans and remove grand vezirs. Before the beginning of nineteenth century their strict discipline had gone and they became a danger to the state itself. By then they were allowed to marry and often sons followed their fathers into the corps. In the Balkans at this time, Muslim guildsmen with no martial prowess were also termed Janissaries. Many of these also became Bektashis, thus acquiring three platforms – trade, military, sectarian – for the expression of grievances.[39]

The deplorable state of the Janissaries became widely known, as this extract from a report sent to Lord Howick, Secretary of State for Foreign Affairs, in January 1807, by Captain W. M. Leake, a British agent, confirms:

The Janissaries of the city [Salonica], who amount to about four thousand, being all natives of the place, and most of them engaged in trade, or other civil occupations, cannot be expected to afford much assistance to the Government in opposing a foreign enemy. But they might perhaps make a very active use of their arms in plundering and massacring the peaceable Rayahs, if the operations of war should approach this quarter of the Empire. It may even be apprehended, that the European merchants, established in the great commercial cities, would not be quite secure from the excesses of the Turkish soldiery, if the Sultan should meet with any alarming reverses, but more particularly, if any of the other Sovereigns of Europe, engaging in the contest, should give the people reason to suspect a concerted design on the part of the Christian Powers to dismember the Turkish Empire.[40]

Clearly, though Janissary connections boosted Bektashi influence, they did not necessarily enhance the order's reputation, and profession of Bektashi faith was no guarantee of a virtuous lifestyle.

It is not known for certain exactly when the Bektashi connection with the Janissaries began. According to the legend that Bektashis relate even today, Haji Bektash was present at their inauguration. When Orhan Gazi (1288–1359) the second Ottoman sultan, who ruled from 1324 to 1359, set up a special military force recruited from prisoners of war, he personally told Haji Bektash to name and bless this new unit. Accordingly Haji Bektash declared, 'Your name shall be *Yeniçeri* [New Force], may your hands be victorious, your swords sharp and your spears ever ready to strike the foe!' As he said these words, he placed his hand on the head of one of these new soldiers. When he did so, the sleeve of his robe hung down over the young man's shoulders. In memory of that incident Janissary headgear thereafter had a flap of material that hung down from its tall crown to the wearer's shoulders.[41] In fact, it does seem likely that Bektashis became associated with the Janissaries at an early stage in their history, though there may have been no formal recognition of this for many years. There was, however, formal recognition around 1591 when the head of the Bektashi order was made an honorary colonel of the Janissaries and eight Bektashi dervishes were appointed to the 99th *orta* (battalion) to perform duties similar to those of an army chaplain. Thereafter, it was the custom for each new head of the Bektashi order to go to Istanbul to be 'crowned' by the commander of the Janissaries.[42]

Although Janissary links had greatly enhanced the growth of Bektashism those same links proved gravely damaging to the order when

Murad II destroyed this troublesome corps on 15 June 1826 – an event that is known in Turkish history as 'The Auspicious Incident'. The previous month, on 28 May 1826, the sultan had in any case issued an imperial decree ordering reforms that would have removed Bektashi influence from the Janissaries, and substituted Sunni orthodoxy.[43] When the Janissaries rebelled against his reforms, the sultan ordered his artillery to bombard them and quell their revolt. Some 4000 Janissaries were killed in the immediate carnage in Istanbul. Thousands more were executed throughout the Empire.[44] Those who were not killed were stripped of their privileges and most could subsequently only find menial employment.[45]

But the sultan and the religious authorities did not restrict their rage to the Janissaries. They directed it at the Bektashis too. Seizing their opportunity, leaders of rival dervish orders and representatives of the orthodox *ulema* sped to denounce Bektashi heresy and infamy. Their reports fell on eager ears. On 10 July 1826[46] the sultan issued a decree abolishing the Bektashi order. Many Bektashi leaders were put to death and many others were sent into exile. A great many went into hiding. Birge records one report that the sultan 'vowed to execute seventy-thousand Bektashis, and that when he could not find that many to behead he ordered headpieces to be cut off Bektashi tombstones until the count should be complete!'[47] The intensity of anti-Bektashi fever reached such a pitch that some people falsely accused their enemies of being Bektashis in order to get them exiled. A number of Bektashis avoided disaster by claiming to be orthodox Sunnis. In so doing they were resorting to *takiye* (Arabic *takiya*),[48] 'the dispensation from the requirements of religion under compulsion or threat of injury'. The Koranic verse 2: 91, 'Throw not yourselves with your own hands into ruin' is held to justify this deception, which Bektashis throughout their history have frequently, indeed almost routinely, felt pressured into.

In addition to the measures against the order's members, the sultan also acted against their property. He had many tekkes razed to the ground. Many others were turned over to the Nakshibendi order and some were turned into mosques. A Nakshibendi representative was installed in the chief Bektashi tekke at Hacibektaş, and a mosque was built within the complex. (This was part of campaign to Sunnify the Bektashis. As we have already noted, Bektashis do not worship in mosques. In Turkey today Bektashis resent the authorities' continued attempts to Sunnify them by building mosques in their villages.)

Bektashis in Turkey 1826–1925

This attempt to wipe out the order did not succeed. Within about twenty-five years they had largely recovered.[49] Their traditional secrecy had helped them weather the storm. That habit of secrecy also made them valuable allies to the Young Turks in the Balkans and elsewhere as they prepared for the Young Turk revolution of 1908. And after Turkey's collapse in the First World War, the Bektashis rendered great assistance to Mustafa Kemal as he led Turkish nationalists during the War of Independence that led to the establishment of the Republic of Turkey in 1923.

Although they enthusiastically welcomed the birth of the Republic, the Bektashis together with all other dervish orders were closed down in 1925 by Mustafa Kemal after a revolt led by a Nakshibendi sheyk had nearly toppled the infant republic. Once more the Bektashi practice of secrecy enabled the order to survive underground in Turkey. Meanwhile, its headquarters and its leader (who was at that time an Albanian *dedebaba*) moved from Hacibektaş to Tirana.

How Bektashism spread to the Balkans

Although the order was officially banned in Turkey in 1925, it continued its activities openly in the Balkans. So, we should now look at how missionaries, settlers and Janissaries had helped Bektashis to become established in the Balkans.

Bektashis laud Sari Saltik[50] and Seyit Ali Sultan as the first missionaries for their order in this region. (In fact they may have been claimed subsequently by the order because they were influential with Ottoman warriors in Rumelia.)[51] They preached in eastern Greece, eastern Bulgaria and southern Romania. Their success is evident from the fact that many of the ideas that were to shape the order in the fourteenth–fifteenth centuries sprang from this area.[52] Sari Saltik deserves special mention because he figures so prominently among the legends that recounted the Bektashis' early missionary activities in the Balkans. Sent by Haji Bektash to Bulgaria, Sari Saltik killed a seven-headed dragon, 'and finally on his death [he] was found in seven coffins, thus accounting for the traditions of his burial in many localities.'[53] In one manifestation he is said to have come to Albania from Corfu before 1400, and founded the Sari Saltik tekke above Kruje, together with six others.[54] He also became identified with Svt. Naum, the Christian healer of Lake Ohrid. Hasluck records being told in 1914 that Sari Saltik, together with a Christian abbot, had miraculously crossed the lake to Ohrid on a straw mat.[55]

Bektashis claim that their dervishes had moved into and settled in western Thrace long before the arrival of Ottoman armies. They mingled with the local Christians and got on well with them, sharing flocks and land, and rendering assistance.[56]

When the Ottoman armies did fight for possession of new territory, the Türkmen warriors and the Janissaries would bring Bektashi influence with them. In the early years of conquest, Janissaries were quartered in captured castles, separate from other troops. But in the course of time, they lived in cities and towns as well. As we have seen, in the years of Ottoman decline, many Muslim guildsmen were enrolled as Janissaries and most of these also joined the Bektashi dervish lodges.[57]

To work the land and increase security in captured territory, the Ottomans brought many thousands of Turks from Anatolia and settled them along strategic routes and in areas where resistance to Turkish rule was strongest. Türkmen tribesmen were in the vanguard of these Ottoman settlers. Many of them felt comfortable with the Bektashi religion. They found it colourful and social. It incorporated pre-Islamic elements familiar to their forebears in Central Asia. It allowed men and women to mix freely and it did not seek to introduce irksome restrictions into every aspect of life. It therefore seemed much more attractive to them than Sunni Islam. Likewise, it was quite readily acceptable to local inhabitants whose own folkloric religious practices were not too dissimilar.[58]

Where Bektashis were sufficiently numerous they built tekkes. In the early days of Ottoman expansion they often received state aid in building these. Murad I, Murad II and Beyazit II all supported tekke building for the dervish orders because these would form the nucleus of new settlements around which agriculture would develop and the area would become more closely knitted into the Ottoman fabric.[59]

Some of the tekkes were imposing buildings in dramatic locations. Others were small, comparatively humble meeting places. Hasluck devotes many pages[60] to listing Bektashi tekkes in Thrace, Bulgaria, Rumania, Serbia, Greece, Macedonia, Bosnia, and Albania, noting that 'the farthest outpost of Bektashism is the *tekke* of Gül Baba [in Budapest], a relic of the Turkish occupation, which is still one of the minor sights of the Hungarian capital.'

Where Bektashis were in contact with local Christians a degree of cross-fertilisation of faith occurred. It was nothing new for Bektashism to absorb elements of other faiths; syncretism had always been one of its characteristics. But now, while some Christians actually became

Bektashis, others simply adopted Bektashi saints and shrines as their own. There were numerous instances of shared saints and festivals. St George and St Theodore both became equated with Khidr Elias, St Nicholas with Sari Saltik, St Charalambos with Haji Bektash, and so forth.[61]

The Turkish conquerors did not force Islam upon their new Balkan subjects. Forced conversion is, in any case, contrary to Islamic teaching. Nor did they simply impose ethnic Turkish rule in the Empire. They adopted and adapted existing systems and by such means as the *devshirme* and the *millet* system they recruited locals into the ruling establishment. Under their *millet* system the Ottomans were usually tolerant of and largely indifferent to the activities of their non-Muslim subjects and delegated considerable authority to the heads of the main religiously defined communities – originally Greek Orthodox, Jewish and Armenian – to run their own internal affairs. They did levy a poll tax (known as *cizye*) on non-Muslim heads of households. The loss of revenue to state coffers resulting from Christian conversion to Islam was thus a further reason why the Ottomans displayed no proselytising zeal. Thus their policies aimed to favour Muslims without persecuting Christians. Nevertheless, there were benefits to be gained by conversion to Islam. Apart from freedom from the poll-tax, the new convert immediately became the equal of any other Ottoman subject and could rise in the state apparatus to the highest level his abilities would allow, since all posts, bar that of sultan, were open to him. Indeed many who went on to become grand vezirs had started life born into Christian families. So, as Temperley elegantly phrases it: 'The foundations of belief were sapped by the pickaxe of interest and the spade of political advantage.'[62]

Of course, as Sunnis, the Ottoman authorities would not have been seeking to proselytise for the Bektashis. Where the Bektashis would have gained members from Christian families is through the *devshirme* system which supplied recruits for the Janissaries.

The paucity of documentary evidence on the subject makes it difficult to present with any assurance a fully detailed picture of conversion to Bektashism in the Balkans. The first concern of Ottoman historians was to record the achievements of the high and mighty. So the deeds of Bektashis, who more usually associated with the underprivileged, were often ignored in the chronicles. However, it seems safe to concur with Professor Norris's judgement that Bektashism in the Balkans generally took root 'peacefully, slowly and without serious opposition'.[63] Occasionally when a Bektashi was in a position of influence, he might favour the interests of the order, as Ali Pasha of Tepelene demonstrated.[64] More often, however, Bektashi involvement in rebellions led to the order being regarded with suspicion and its members being persecuted. And strict Sunnis, regarding Bektashis as heretics, would persecute them with even more righteous zeal.

As elsewhere in the empire, Bektashi fortunes in the Balkans suffered a severe blow when the order was suppressed after 'The Auspicious Incident' in 1826. Very many Balkan tekkes were destroyed or handed over to rival orders and Bektashi babas were persecuted. The Bektashis' relationship with the rebellious Ali Pasha of Tepelene, who was eventually defeated in 1822, gave the sultan added reason to seek their destruction. The order recovered towards the end of the century but with the outbreak of the Balkan Wars in 1912 and the resultant Turkish withdrawal from the region its numbers sharply declined.

Vestiges of Bektashism remain in the Balkans today. Leaving aside the situation in Albania which will be covered separately, in the rest of the region, some evidence is still to be found. Although many tekkes have been completely destroyed and exist now only in the memory of local inhabitants, a few still survive. In 1976, Dr Nimetullah Hafiz, from Yugoslavia, in a paper delivered at a conference in Hacibektaş, noted reports of former tekkes in Sarajevo, Mostar, Tuzla, Izvornik, Gradishka, Konyich, Skopje, Monasstir (Bitola), Kalkandelen (Tetovo), Ishtip, Tito Veles (Koprulu) Kicevo Debar, Kanatlar, Prizren and Djakova. But he said he knew of only three still functioning at that time in the whole of the then state of Yugoslavia.[65]

Among the former tekkes he mentioned, the Sersem Ali Baba, in Kalkandelen (Tetovo), Macedonia (also known as the Harabati Baba tekke; both names commemorated babas who had once been in charge of it) was, and remains, a most impressive complex of great architectural and cultural significance. It had been one of the largest Bektashi tekkes in the Balkans. However, because of a disputed inheritance, it ceased to be used as a tekke in late Ottoman times. In President Tito's time its potential was recognised and it was preserved and used as a prestigious restaurant. In Kiçevo, a fine, large tekke had been demolished in Ottoman times, but a small tekke was later built to replace it. In 1979 Baba Ziya Pasholi was ministering there to a faithful Bektashi community that continued to celebrate their traditional festivals, including picnics in the countryside at Hidrellez (6 May, regarded as the start of summer). At Kanatlar a house was being used for Bektashi gatherings. The Prizren tekke was closed in 1924, but in 1976 still stood as a private house. In Djakova (Dakovica, Gjakova), a new Bektashi tekke was built by a baba who had previously belonged to another order. (When I visited it in

1979, this tekke, sometimes known as the Şemsettin Baba tekke after its founder, was in the charge of Qazim Baba, an illustrious and greatly revered Bektashi with a long lifetime of service to the order.) This was destroyed in the 1999 conflict.

Some Bektashis survive in western Thrace. They still recall with pride the distinguished Bektashis who came from this region, foremost of whom was, of course, the order's second founder, Balim Sultan, who had been in charge of the tekke at Dimetoka. Another historical figure from this region whom all Bektashis honour is the distinguished scholar and celebrated rebel, Şeyh Bedreddin (1368–1420). He was born in Simavni, close to Dimitoka. His rebellion was brutally suppressed and he was hanged in Serres in 1420, but his concept of justice inspired generations of Bektashis. The poet Nazim Hikmet wrote a moving epic about him which helped to make him a potent symbol to left-wing activists in Turkey during the political turmoil of the 1970s.

Abdurrahim Dede, seeking material evidence of Bektashi presence, researched old lists of Bektashi buildings – tekkes and zaviyes (small dervish lodges) – in Thrace, visited the sites and found around a hundred still in existence, including Balim Sultan's tekke in the village of Mikroderion (Demirviran) near Dimetoka, though this was in a prohibited area.[66] He also found that the Greeks had used 'roadworks' as a pretext to demolish several historic tekkes.

In 1974, in response to a request from Professor Machiel Kiel, who has spent very many years painstakingly researching Bektashi and other Islamic traces in the Balkans, the Romanian government restored a fifteenth-century Bektashi türbe at Babadag. Professor Kiel has found that no trace now remains of many of the fine Bektashi buildings once so common in the Balkans. Several of those destroyed after the 1826 'Auspicious Incident' were rebuilt later in the century as simpler and humbler constructions, but in the 1912–13 Balkan Wars most were again demolished. After the exchange of populations agreed at Lausanne in 1923 had removed Muslims from Greek Macedonia and Epirus, all traces of tekkes there were destroyed. However, in the 1970s it was possible to see remains, in greater or lesser degree, of some important examples. These included the Akyazili tekke, which was built around 1500 with financial assistance from the famous Mihaologlu family,[67] the Demir Baba tekke near Kemanlar, Otman Baba near Haskova, Ali Baba near Kirçaali and the Kidemli Baba tekke near Nova Zagara. According to Evliya Çelebi, the Kidemli Baba tekke, one of the biggest in Thrace, was built on the orders of Sultan Mehmet I, who reigned 1413–21, and was the probably the work of Haji Ivas Pasha, who was also the architect of the famous Green Mosque in Bursa. The Demir Baba tekke was a sixteenth-century work, and the Otman Baba tekke seems to belong to the reign of Mehmet II, which ended in 1481.[68]

Despite strong Janissary influence in Bosnia – so strong that Mahmud II had sent a special expedition to suppress them five years after the 'Auspicious Incident' because they had been protected by local notables – it would appear that Bektashism never achieved similar importance there.[69]

Bektashism in Albania

The most fertile soil for Bektashism outside Anatolia proved to be Albanian. In Albania, by demonstrating adaptability, the order achieved greater influence and survived the break up of the Ottoman Empire more successfully than in the rest of the Balkans. One of the order's enduring characteristics has been its capacity to adapt to local conditions and appeal to local people and national sentiment. In Turkey Bektashi poets had produced some fine literature in pure Turkish comprehensible to the ordinary people instead of using the rarefied Ottoman that was the preserve of the élite. In Albania, too, Bektashis made valuable contributions to their national literature. In modern Turkey Bektashis expediently stress the order's Turkishness, even to the extent of publicising an apocryphal saying of Haji Bektash, 'The Turkish people were created to rule the world'.[70] Similarly, in Albania, Bektashis expressed Albanian sentiments. Writing in 1983, Philip Ward accurately summarised their position as follows: 'the Bektashi movement...included so many Albanian nationalists that it can fairly be described as the most representative nationalist movement in Albania. While Orthodoxy was led to Greek colonial aspirations, Catholicism to the Vatican, traditional Islam to Turkey, and the Party of Labour now in power to international communism, Bektashism in its peculiar Albanian form undermined all factions and opposites, mixed pagan, Christian and Muslim elements, and stood for mystic unity, intellectual honesty, and universal tolerance.'[71] The great national hero Skanderbeg (1403–68) is said to have been a Bektashi while in Ottoman service. Ali Pasha of Tepelene (1741–1822) became a Bektashi and whilst ruling Albania with a considerable degree of independence he supported the spread of Bektashism. He was reportedly initiated by Shemimi, who won many converts and opened numerous tekkes.[72] Bektashi allegiance did not, however, make Ali Pasha an exemplar of the order's virtues![73]

Albania, the majority of whose population was Muslim, remained loyal to the Ottomans longer than other Balkan countries. After the Congress of Berlin (1878) Albanians felt a sense of national solidarity and their desire for independence became evident. At this time Ottoman education was organised according to the *millet* system. Muslim Albanians were taught Turkish and Arabic, but Christians there belonged to the Orthodox *millet* and were taught in Greek. This Turkish failure to recognise Albanian language or culture caused resentment that fuelled nationalist feeling. In 1880–1 a national movement arose, and this was organised through Bektashi tekkes.[74] Abdul Bey Frashëri intended to use this movement to gain Albanian independence. His brother, Naim Bey, in a 32–page pamphlet that he wrote primarily to expound the order's beliefs and from which we have already quoted, gave an Albanian slant to Bektashism as is clearly shown in this further extract:

> Let them strive night and day for the nation to which the Father calls them and vouches for them that they will work with the chiefs and the notables for salvation of Albania and the Albanians, for the education and civilization of their country, for their language, and for all progress and improvement.
>
> Together with the chiefs and notables let them encourage love, brotherhood, unity, and friendship among all Albanians: let not the Mussulmans be divided from the Christians, and the Christians from the Mussulmans, but let both work together. Let them strain towards enlightenment, that the Albanian, who was once reputed throughout all the world, be not despised today.[75]

Bektashi disillusionment with Ottoman rule intensified when the 1908 Young Turk revolution, in which many Bektashis had been involved, resulted not in a greater degree of independence for Albania but in attempts to control its affairs more firmly from Istanbul.[76] Albanian nationalists began a revolt in 1910[77] which led to the declaration of independence in 1912.

Bektashism subsequently received recognition as one of the country's religions and its leader was granted an *ex-officio* seat in the legislative National Assembly in 1914.[78] The patriotic pride of Albanian Bektashis was evident at their pan-Bektashi Congress held at Berat in January 1921 when they declared themselves to be the first religious body in the country to be free from all foreign domination.[79] The next year the Albanian Bektashi assembly broke away from the order's headquarters in Turkey.[80] In 1925, when the Bektashis and other dervish orders were suppressed in the Turkish Republic, the Supreme Bektashi, an Albanian, settled at Tirana.[81] In 1929 Bektashism was recognised as autonomous within the Muslim community of Albania, with statutes drawn up at Korçe.[82] In 1930 the government approved the regulations under which Bektashi affairs were regulated.[83] Birge, writing of the situation around that time, estimated the number of Bektashis in Albania as 150000 to 200000, and reported:

> The whole country is divided into six dioceses, Prishte, Kruja, Elbasan, Korcha, Frasheri and Gjinokaster. Government is through Local Councils and a mixed Council of twelve members elected by secret ballot, two from each diocese, one being a Father or Grandfather, the other an Initiated or Confirmed Member. Another Assembly called the Holy Council of Grandfathers is made up of the diocesan heads with the Arch-grandfather as chairman. The Regulations provide for the possibility of a theological seminary for the training both of clergy and of candidates for confirmation.
>
> Many of the leaders in Albania today are Albanian Bektashis who had their training in Turkey under the old regime, and who have now returned to their fatherland in positions of influence. Not as numerous as the Sunni Moslems, the Bektashi community nevertheless constitutes some fifteen to twenty per cent of the total population of the country, and is recognized by all as one of the worthiest elements in the population.[84]

(The terms Arch-grandfather, Grandfather and Father correspond to the *Dedebaba, Halife,* and *Baba* mentioned earlier.)

From 1935 to 1936 Albania had a Bektashi prime minister, Mehdi Frashëri, a liberal idealist. He later became President of the Council of Regency under Italian administration in the Second World War.[85]

In occupied Albania in the Second World War, the Bektashi leader, Niazi Dede, was murdered in January 1942. Thereafter many Bektashis were active in the guerrilla movement.[86] David Smiley, a Special Operations Executive agent in Albania during the war, gives a graphic account (as well as a photograph) of one of them, Baba Faja:

> His real name was Baba Mustafa, though he was always known as Baba Faja. A priest or *hoj* of the Bektashi sect. His monastery at Martenesh had been burnt down some time before by the Italians. He was a well-built and rather stout man, with a massive black beard.

Apart from his priest's hat, which he always wore, his usual dress was a loud check plus-four suit over which he slung his bandolier and pistol. He was a likeable character, but a scoundrel, and he drank heavily. (Bektashis are in fact permitted to drink alcohol, unlike other Moslem sects.) He delighted in singing partisan songs in his deep bass voice, especially after consuming large quantities of raki. We did not take long to discover he was being used as a figurehead by the partisans, and always had a commissar at his elbow to keep an eye on him.[87]

Smiley also reports Baba Faja's end:

Baba Faja is dead, killed by the Supreme Head of the Bektashi Sect, Dede Baba Abazi. It appears that a religious discussion between these two and another priest turned into a heated argument; Baba Abazi drew his pistol and shot both Baba Faja and the other priest, and then turned the pistol on himself and committed suicide.[88]

After the war, the communist regime reportedly reorganised the Bektashi community.[89] With the headquarters and world leader still in Tirana, six areas were established, each under what in Turkish would be called a *halife*:

Area	Main tekke
Krujë/Durres/Shkodër	Fushë Krujë
Elbesan/Peshkopi	Krastë
Korçe	Melcon
Gjirocastra	Asim Baba
Berat/Permet	Prishtë
Vlorë	Frashër's

Persecution of Bektashis in Albania apparently began in the early 1960s, before the major anti-religious campaign was launched in 1967.[90] In atheist Albania, most of the 300-plus tekkes were closed and those that were not destroyed were put to other purposes.[91] The library of the chief tekke in Tirana was burnt and the tekke itself was, according to Vickers and Pettifer[92] turned into an old people's home. (A report in the Turkish newspaper *Zaman* said it was used as a brothel.)

Some Bektashi babas were sentenced to forced labour on prison farms. Tekkes of architectural importance were preserved and by the late 1970s some restoration work was carried out, as the authorities became aware of their value as tourist attractions. (In July 1987 I was shown restoration work in progress on the small Bektashi tekke called Dollma in Kruja.)[93]

Bektashis re-emerged after the fall of hard-line communists in Albania in 1990. They held their first formal assembly in June the same year.[94] With financial aid from Bektashi compatriots abroad, they began renovating their tekke in Tirana. Its formal reopening was attended by Mother Theresa and other distinguished guests.[95] The current head of this *tekke* is Baba Reshat Bardhi. Under Enver Hoxha, who had himself been born into a Bektashi family,[96] he had served a long term of forced labour on a prison farm in southern Albania. By 1996 a new site for the chief tekke was being developed near the Botanic Gardens in Tirana.[97] Vickers and Pettifer reported that, 'other important *tekkes* have reopened, such as the exceptionally beautiful building in the citadel at Kruja, and a new one has been established in the eastern highlands at Fushe Bulquiza'. They did not, however, consider that Bektashism would regain its former importance in the country; its hold on the young was weak.[98]

Materialism has captured the hearts of many Albanians since the communist yoke has been removed, so there are fewer now who would accept the sentiment expressed in the account of Ahmad Sirri Baba's life: 'He who is discerning and bright seeks perfection, while he who is ignorant seeks money.'[99]

International contacts

In recent years, with the relaxation of international travel restrictions and the increasing number of people seeking employment outside their own countries, Bektashis have become more widely dispersed throughout the world. At the same time, they have also found it easier to gather together. One focus for gatherings is the town of Hacibektaş. There, in mid-August an annual Haji Bektash Festival is now held. Officially, since Law No. 677 that banned the dervish orders in Turkey in 1925 has never been rescinded, this is a 'cultural' rather than a religious event. In practice the occasion demonstrates that Bektashism has survived that suppression as it survived previous attempts to destroy it. Bektashis come from every continent where they happen to be living to join the celebrations. While some attend only the public folkloric activities, those Bektashis with a serious commitment to their faith visit the tekke, which was reopened in the 1960s as a museum, and other sites, including the legendary cave and rocks that feature in the traditional

Bektashi pilgrimage, and also gather behind closed doors for secret ceremonies and ritual meals complete with traditional candles. Hymns are sung there in praise of Ali and his family and curses uttered against his enemies, particularly the first three caliphs.

For many years Bektashis from the Balkans have been welcome guests at these gatherings. They include a number who migrated from Yugoslavia to Turkey when Tito's government allowed them to leave. Some of these settled in Istanbul, others in Turgutlu, integrating harmoniously with Turkish Bektashis. Others make the journey from various parts of the Balkans, including now Albania. Yet others come from western Europe, where Bektashis among the *Gastarbeiter* have in recent years made vigorous efforts to assert their identity and publicise their distinctive culture by establishing tekkes and organising gatherings.

Figure 11.1 A Bektashi tombstone at Kruja showing the traditional Bektashi headgear with twelve symbolic ridges.

Those attending the three-day festival at Hacibektaş vary widely in their convictions. There are fewer now who believe all the old legends with their accounts of amazing miracles, but even the sceptics among them still treasure 'those illusions that give such colour to the world that you don't care whether things are true or false as long as they partake of the magical glory'.[100]

Notes

1 See L. Massignon, 'Tasawwuf' in *Shorter Encyclopaedia of Islam*, p. 580, and Fazlur Rahman, *Islam*, p. 156.
2 See L. Massignon, 'Tarika' and 'Tasawwuf' in *Shorter Encyclopaedia of Islam*.
3 Alfred Guillaume, *Islam*, p. 141.
4 Halil Inalcik, *The Ottoman Empire: The Classical Age 1300–1600*, p. 194.
5 See J. K. Birge, *The Bektashi Order of Dervishes*, pp. 34–5.
6 See I. Mélikoff, 'Bektashi / Kizilbaş: Historical Bipartition and Its Consequences' in Tord Olsson et al. (eds), *Alevi Identity*, pp. 1–2.
7 B. Noyan, *Bektaşîlik Alevîlik Nedir*, p. 19, and A. C. Ulusoy, *Hünkâr Haci Bektaş Veli ve Alevî-Bektaşî Yolu*, p. 19.
8 Halil Inalcik, *The Ottoman Empire: the Classical Age 1300–1600*, pp. 191–4.
9 Ahmet Yaşar Ocak, 'Bektaşilik' in Türkiye Diyanet Vakfi, *İslam Ansiklopedisi*, Cilt 5, Istanbul, pp. 373–9.
10 See J. K. Birge, *The Bektashi Order of Dervishes*, pp. 132–4.
11 See Birge, pp. 132–4, 145–6, F. W. Hasluck, *Christianity and Islam under the Sultans*, p. 554, and Ulusoy, pp. 173–5.
12 Noyan, p. 54.
13 Noyan, pp. 396–7; Ulusoy, pp. 112–18, 187–9, 202–3, 240–7.
14 Noyan, pp. 65–6. Virani Baba, writing over a century ago, expressed the same teaching. See Birge, pp. 96–7.
15 Noyan, pp. 65 and 67.
16 See, for example, Ulusoy p. 258, and the novel *Nur Baba* by Yakup Kadriosmanoğlu.
17 Birge, pp. 106–7.
18 Birge, p. 83. But see also Inalcik, p. 199. He considers Evliya Çelebi's estimate of a total of 700 Bektashi tekkes may be an exaggeration.
19 Noyan, pp. 275–89. It might be noted that within the Sufi brotherhoods in the Balkans, as elsewhere in the World of Islam, there is a considerable overlap between them. Some Sufis are affiliated to more than one order. Loyalties and babas are often shared and *tekkes* used jointly. For example, in Orahovac, in Kosovo, a *tekke*, which was badly damaged during the recent conflict and its baba reportedly murdered, was shared between the *Khalwatiyya* and the *Rifaiyya*. Many other examples of this may be found. For example, the Bulgarian *tekkes*, which have been mentioned, such as Akyazili, near Varna, one or two near Haskova, and above all, Demir Baba, are claimed by the *Babaiyya*, they are quite distinct from the *Bektashiyya*, despite shared adoration for Ali and other great men of Islam. These *tekkes*, and others, are the 'holy of holies' of the Bulgarian Alawite Kizilbaş. On the distinction between 'Bektashiyya' and 'Babaiyya', *see* Irène Mélikoff's definitive volume, *Hadji*

Bektach, Un Mythe et ses Avatars, Brill: Leiden; Boston: Koln, 1998, pp. 146–78 and especially pp. 148–9.

20 Birge, p. 165.
21 For details see Noyan, pp. 292–6, and Birge, pp. 170–1.
22 Ulusoy, p. 167.
23 Noyan, p. 61.
24 Some Bektashis do fast in Ramadan. In 1979 Qazim Baba at Djakova told me he did.
25 Najaf and Kerbala in Iraq have also attracted Bektashi pilgrims (a Bektashi tekke was built in each of these locations). It is not entirely unknown for Bektashis to make the pilgrimage to Mecca. For example, an account of the life of Shaykh Muhammad Lutfi Baba in H. T. Norris, *Islam in the Balkans,* pp. 218–21, notes his pilgrimage there at the age of fifty. The same account states that this Bektashi 'followed the *Sharia* and the *Sunna* of the Prophet'. He is also particularly unusual for having built a *mosque.* This was in Gyrokaster, his birthplace.
26 H. T. Norris, *Islam in the Balkans,* Hurst & Company, London, 1993, p. 165.
27 Noyan, p. 297.
28 Mehmet Eröz, *Türkiye'de Alevilik Bektaşilik,* 1997, Istanbul, p. 398.
29 H. T. Norris, *Islam in the Balkans,* Hurst & Company, London, 1993, p. 90. This excellent book contains illustrations and abundant information about Bektashism in the Balkans.
30 See Birge, p. 271 and Noyan, pp. 54–9.
31 See Eröz, pp. 184–8.
32 Hasluck, pp. 554–62.
33 Stanford J. Shaw, *History of the Ottoman Empire,* Vol. 1, p. 26.
34 R. J. Crampton, *A Concise History of Bulgaria,* p. 33.
35 Ottaviano Bon, *The Sultan's Seraglio,* London, Saqi Books, 1996, pp. 59–60.
36 See Shaw, pp. 113–14, Crampton, pp. 33–4, and Godfrey Goodwin, *The Janissaries,* pp. 32–53.
37 Birge, pp. 74–8, Goodwin, p. 30, H. Inalcik, 'The Rise of the Ottoman Empire', in M. A. Cook (ed.), *A History of the Ottoman Empire to 1730,* p. 47.
38 Crampton, p. 34.
39 See A. N. Kurat and J. S. Bromley, 'The Retreat of the Turks, 1683–1730' in M. A. Cook (ed.), *A History of the Ottoman Empire to 1730,* p. 182.
40 H. W. V. Temperley, *History of Serbia,* p. 334.
41 Ali Sümer, *Anadoluda Türk Öncüsü Haci Bektaş Velî,* p. 11.
42 Goodwin, p. 148, and Birge, pp. 74–8.
43 Goodwin, p. 217.
44 Birge, p. 77.
45 Goodwin, p. 230.
46 Goodwin, p. 231.
47 Birge, p. 78.
48 R. Strothmann, 'Takiya', in *Shorter Encyclopaedia of Islam,* p. 561, and Birge, pp. 78 and 270.
49 Birge, p. 79.
50 H. T. Norris, *Islam in the Balkans,* provides extensive information about this legendary character and his importance in the Balkans. See especially pp. 132–3, 146–55. Professor Norris's book also contains a most valuable

discussion of how Islam in general and Bektashism in particular spread in the Balkans.

51 See Inalcik, *The Ottoman Empire: The Classical Age 133–1600*, p. 194.
52 Machiel Kiel, 'Güney Romanya'da Sari Saltik'in Çalişmalari ve Doğu Bulgaristan'da Erken Bektaşilik Üzerine Tarihsel Önem Taşiyan Notlar', in *Haci Bektaş Veli: Bildiriler, Denemeler, Açikoturum*, p. 14.
53 Birge, pp. 51–2.
54 [Admiralty] Naval Intelligence Division, *Albania*, p. 146.
55 Hasluck, pp. 436 and 583.
56 Abdurrahman Dede, 'Bati Trakya'da Bektaşîlik ve Bektaşîlik Hakkinda Arşiv ve Kütüphane-lerimizde Bulunan Yazma Eserler', in *Haci Bektaş Veli: Bildiriler, Denemeler, Açikoturum*, p. 45.
57 See A. N. Kurat and J. S. Bromley, 'The Retreat of the Turks, 1683–1730', in M. A. Cook (ed.), *A History of the Ottoman Empire to 1730*, p. 182.
58 See Peter F. Sugar, *Southeastern Europe under Ottoman Rule, 1354–1804*, Seattle, University of Washington Press, 1977, pp. 53–4, quoted in H. T. Norris, *Islam in the Balkans*, pp. 264–5.
59 Karen Barkey, *Bandits and Bureaucrats: the Ottoman Route to State Centralization*, Ithaca and London, Cornell University Press, 1994, pp. 125–6.
60 Hasluck, pp. 518–51.
61 See Shaw, p. 28, Birge, p. 39 Note 3, and Hasluck, pp. 83–4.
62 Temperley, p. 111.
63 H. T. Norris, *Islam in the Balkans*, p. 124.
64 See H. T. Norris, *Islam in the Balkans*, pp. 239–40.
65 Nimetullah Hafiz, 'Yugoslavya'da Bektaşî Tekkeleri', in *Haci Bektaş Veli: Bildiriler, Denemeler, Açikoturum*, pp. 33–7.
66 The full list is given on pp. 48–9 of the paper he delivered at a conference in Hacibektaş in 1976 and is in the published proceedings: *Haci Bektaş Veli: Bildiriler, Denemeler, Açikoturum*, Ankara, 1977.
67 Professor Norris has drawn attention to the seven-sided plan of this building, since it illustrates Bektashi interest in numerology. Harry T. Norris, 'Aspects of the Sufism of the Rifa'iyya in Bulgaria and Macedonia in the Light of the Study Published by Canon W. H. Y. Gardiner in the *Moslem World*', in *Islam and Christian-Muslim Relations*, Vol. 7, No. 3, 1996, pp. 297–309, p. 304. In his book *Islam in the Balkans*, Norris also refers, among others, to tekkes in Qatrina (p. 223), Monastir (p. 51), Belgrade (p. 123), Farsala (p, 129) and Darfalli (p. 135).
68 Machiel Kiel, 'Güney Romanya'da Sari Saltik'in Çalişmalari ve Dogu Bulgaristan'da Erken Bektaşilik Merkezi Üzerine Tarhisel Önem Taşiyan Notlar', pp. 13–29.
69 See Erik Cornell, 'On Bektashism in Bosnia', in Tord Olsson et al. (eds), *Alevi Identity*, Istanbul, 1998, pp. 9–12.
70 Sümer, p. 27.
71 Philip Ward, *Albania*, Cambridge, Oleander Press, 1983, pp. 102–3.
72 Birge, p. 72.
73 For further comments on the nature of Ali Pasha's devotion to Bektashism see H. T. Norris, *Islam in the Balkans*, pp. 239–40.
74 Hasluck, p. 552.
75 Hasluck, p. 562.

76 Shaw, pp. 199–200.
77 Shaw, p. 288.
78 Admiralty, pp. 208.
79 Primrose Peacock, p. 39.
80 Admiralty, p. 146.
81 Albanians were usually prominent in the chief tekke in Hacibektaş. H. T. Norris, *Islam in the Balkans*, pp. 91–2 quotes from an article written by G. E. White in 1913 in which he reports finding around 80 Albanian babas there.
82 Admiralty, p. 146.
83 Birge, p. 85 referring to *Rreegullore e Bektashijvet Shqiptare*, Tirane, 1930.
84 Birge, p. 86.
85 Admiralty, p. 194.
86 Admiralty, p. 230.
87 David Smiley, *Albanian Assignment*, London, Sphere Books, 1984, p. 84.
88 Smiley, pp. 159–60.
89 Peacock, p. 40.
90 Miranda Vickers and James Pettifer, *Albania: From Anarchy to a Balkan Identity*, New York, New York University Press, 1997, p. 101.
91 Vickers and Pettifer, p. 99.
92 Vickers and Pettifer, p. 101.
93 This work continued, as Norris, in *Islam in the Balkans*, refers to restoration in 1989, p. 128.
94 Vickers and Pettifer, p. 101.
95 In view of the anti-Bektashi stance of orthodox Muslims, it is mildly amusing to note that this reopening was hailed in Turkey by the staunchly Sunni newspaper *Zaman* as a triumph for Islam.
96 Jon Halliday (ed.), *The Artful Albanian: the Memoirs of Enver Hoxha*, London, Chatto & Windus, 1986, p. 2.
97 Peacock, p. 30.
98 Vickers and Pettifer, pp. 101–2.
99 H. T. Norris, *Islam in the Balkans*, p. 221.
100 As F. Scott Fitzgerald said of *The Great Gatsby*.

References

[Admiralty] Naval Intelligence Division, *Albania* (London, 1945).
Barkey, Karen, *Bandits and Bureaucrats: the Ottoman Route to State Centralization* (Ithaca and London: Cornell University Press, 1994).
Birge, John Kingsley, *The Bektashi Order of Dervishes* (London: Luzac & Co Ltd, 1965).
Bon, Ottaviano, *The Sultan's Seraglio* (London: Saqi Books, 1996).
Cornell, Erik, 'On Bektashism in Bosnia' in Tord Olsson et al. (eds), *Alevi Identity* (Swedish Research Institute in Istanbul, 1998).
Crampton, R. J., *A Concise History of Bulgaria* (Cambridge: Cambridge University Press, 1997).
Dede, Abdurrahman, 'Bati Trakya'da Bektaşîlik ve Bektaşîlik Hakkinda Arşiv ve Kütüphanelerimizde Bulunan Yazma Eserler', in *Haci Bektaş Veli: Bildiriler, Denemeler, Açikoturum* (Ankara: Hacibektaş Turizm Derneği Yayinlari, 1977).
Eröz, Mehmet, *Türkiye'de Alevilik Bektaşilik* (Istanbul, 1997).

Goodwin, Godfrey, *The Janissaries* (London: Saqi Books, 1994).

Guillaume, Alfred, *Islam* (Harmondsworth: Penguin Books, 1954).

Hafiz, Nimetullah, 'Yugoslavya'da Bektaşî Tekkeleri' in *Haci Bektaş Veli: Bildiriler, Denemeler, Açikoturum* (Ankara: Hacibektaş Turizm Dernegi Yayinlari, 1977).

Halliday, Jon (ed.), *The Artful Albanian: the Memoirs of Enver Hoxha* (London: Chatto & Windus, 1986).

Hasluck, F. W., *Christianity and Islam under the Sultans* (Oxford: Oxford University Press, 1929).

Inalcik, Halil, *The Ottoman Empire: the Classical Age 1300–1600* (London: Weidenfeld and Nicolson, 1973).

Inalcik, Halil, 'The Rise of the Ottoman Empire', in M. A. Cook (ed.), *A History of the Ottoman Empire to 1730* (Cambridge: Cambridge University Press, 1976).

Kiel, Machiel, 'Güney Romanya'da Sari Saltik'in Çalişmalari ve Doğu Bulgaristan'da Erken Bektaşilik Üzerine Tarihsel Önem Taşiyan Notlar', in *Haci Bektaş Veli: Bildiriler, Denemeler, Açikoturum* (Ankara: Hacibektaş Turizm Dernegi Yayinlari, 1977).

Kurat, A. N. and Bromley, J. S., 'The Retreat of the Turks, 1683–1730', in M. A. Cook (ed.), *A History of the Ottoman Empire to 1730* (Cambridge: Cambridge University Press, 1976).

Massignon, L., 'Tarika' and 'Tasawwuf', in *Shorter Encyclopaedia of Islam* (Leiden: E. J. Brill, 1974).

Mélikoff, I., 'Bektashi / Kizilbaş: Historical Bipartition and its Consequences', in Tord Olsson et al (eds), *Alevi Identity* (Swedish Research Institute in Istanbul, 1998).

Norris, Harry T., 'Aspects of the Sufism of the Rifa'iyya in Bulgaria and Macedonia in the Light of the Study Published by Canon W. H. Y. Gardiner in the *Moslem World*', *Islam and Christian-Muslim Relations*, 7(3) 1996.

Norris, H. T., *Islam in the Balkans* (London: Hurst & Company, 1993).

Noyan, Bedri, *Bektaşîlik Alevîlik Nedir* (Ankara, 1987).

Ocak, Ahmet Yaşar, 'Bektaşilik', in Türkiye Diyanet Vakfi, *İslam Ansiklopedisi*, Cilt 5 (Istanbul, 1992).

Rahman, Fazlur, *Islam* (2nd edition) (Chicago: University of Chicago Press, 1979).

Shaw, Stanford J., *History of the Ottoman Empire*, Vol. 1 (Cambridge: Cambridge University Press, 1976).

Smiley, David, *Albanian Assignment* (London: Sphere Books, 1984).

Strothmann, R., 'Takiya' in *Shorter Encyclopaedia of Islam* (Leiden: E. J. Brill, 1974).

Sümer, Ali, *Anadoluda Türk Öncüsü Haci Bektaş Velî* (Ankara, 1970).

Temperley, H. W. V., *History of Serbia* (London: G. Bell and Sons Ltd, 1919).

Ulusoy, A. C., *Hünkâr Haci Bektaş Veli ve Alevî Bektaşî Yolu* (Hacibektaş, 1980).

Vickers, Miranda and Pettifer, James, *Albania: From Anarchy to a Balkan Identity* (New York: New York University Press, 1997).

Ward, Philip, *Albania* (Cambridge: Oleander Press, 1983).

Part IV

Religion, Politics, National Mythologies

12
Religion, Irreligion and Nationalism in the Diaries of the Bulgarian Exarch Yosif

F. A. K. Yasamee

The diaries of the Bulgarian Exarch Yosif are not new to historical scholarship; nonetheless, their recent publication in full is to be welcomed, and not least because historical scholarship to date has obscured their importance.[1] Commenced in 1868, and continued until the Exarch's death in 1915, the diaries fall into that relatively rare category of Balkan historical sources, private papers; they were not written for publication, and despite occasional gaps and omissions – some, perhaps, deliberate – they possess a depth and credibility rarely found in official documents. Their importance is twofold. In the first place, they are an invaluable source of information on the affairs of the Bulgarian Orthodox Church, the Bulgarian Principality, the Ottoman Empire and Macedonia, and much of this information challenges the accounts of these affairs which have been propagated by historians. In the second place, the diaries are much more than a record of events; they also offer a view of those events and of their meaning, a view shaped by the Exarch's own strong religious, political and moral convictions. This view, too, challenges received history.

I

Exarch Yosif was an unusual prelate. Born Lazar Yovchev in 1840, he received a lay education, culminating in three years' study of literature and law at the Sorbonne. Upon his return to the Ottoman Empire in 1870, he participated actively in the movement for Bulgarian cultural and national revival, notably as editor of the periodical *Chitalishte*. In 1872 he entered the service of the newly established Bulgarian

Orthodox Church, or Exarchate, as a monk, and within four years had been elevated to the position of Metropolitan of Lovech – testimony to his own qualities, but also to the dearth of suitably qualified men within the Bulgarian church hierarchy. In May 1877 he was installed as Exarch: head of the Bulgarian Orthodox Church, and also, under the Ottoman millet system, legally recognised spiritual and political leader of the Bulgarian nation.[2] He assumed office at a critical turning-point in the fortunes of the Church and the Bulgarian national movement, immediately after the outbreak of a war between the Ottoman Empire and Russia, a war provoked in part by Bulgarian revolts against Ottoman rule. The war ended in an Ottoman defeat, leading to the creation, in 1878, of two Bulgarian statelets, nominally dependent on the Ottoman Sultan: a principality of Bulgaria headed by the German Protestant Prince Alexander Battenberg, with its capital at Sofia, and an autonomous province of Eastern Rumelia, with its capital at Plovdiv. Macedonia and southern Thrace remained under direct Ottoman rule, leaving the Exarchate's jurisdiction divided between three separate political authorities.[3]

Exarch Yosif had a clear view of the task that awaited him. He was a nationalist, who saw the Exarchate as a vehicle for the dissemination and affirmation of Bulgarian national consciousness.[4] The Exarchate, established by the Ottoman government in 1870 after a protracted struggle between Bulgarian nationalists and the Oecumenical Orthodox Patriarchate of Constantinople, had been the Bulgarians' first national institution, with a jurisdiction extending over almost the whole of the future territories of the Bulgarian principality and Eastern Rumelia; within these regions, it had controlled their churches, schools and local community organisations; and in establishing it, the Ottoman government had provided for the possibility of future extensions of its jurisdiction into Macedonia and southern Thrace, where the development of Bulgarian national consciousness was as yet laggardly. Already, by 1877, the Exarchate had increased its Macedonian eparchies to three, at Veles, Ohrid and Skopje, and further eparchies were confidently expected.[5] As of 1878, the success of the Bulgarian national movement in Bulgaria and Eastern Rumelia was assured; Exarch Yosif proposed to focus the Church's national mission on those regions which remained under direct Ottoman rule: 'the Exarchate remains for the sake of Macedonia'.[6]

Not that Exarch Yosif's view of religion was purely instrumental: he was a sincere Orthodox Christian, who saw Orthodoxy not only as a defining element of Bulgarian national identity, but also as an irreplace-

able source of individual and communal values crucial to the development of a modern Bulgarian nation. In his view, the national revival was
also a religious revival, which had enabled Bulgarians to hear God's
Word in their own tongue, and must continue as such. The Church,
but newly founded, was poor, and its clergy were largely ill-educated; it
must be elevated materially and spiritually, so that it might fulfil its task
of infusing Orthodox, Christian values into the emerging Bulgarian
nation. Time would be required. Exarch Yosif was a conservative and
an evolutionist: he believed that the Bulgarians, an overwhelmingly
illiterate, peasant people, would require a lengthy period of cultural,
social and economic development before they could safely assert themselves as a fully independent nation. Until then, the Bulgarians must
remain politically loyal to their suzerain, the Ottoman Sultan, and to
their liberator, Russia. Exarch Yosif was not a reactionary: he acknow-
leged the European liberal ideal of the nation as a community of free
and equal citizens. He also held, however, that such a community
became feasible only when its members had reached an appropriate
level of intellectual and moral development. In this respect, he insisted,
the Bulgarians still had far to travel: their national consciousness was a
recent and as yet superficial growth, in the mass they were illiterate and
ignorant, their sense of social responsibility remained parochial, and
they were politically immature.[7]

Wise political leadership was essential, for the Bulgarians faced
numerous enemies: the Oecumenical Orthodox Patriarchate, which
had refused to recognise the Exarchate, denouncing it as schismatic,
and insisting that it must either renounce its aspirations to jurisdiction
in Macedonia and southern Thrace, or else renounce Orthodoxy; the
Greek nationalists associated with the Patriarchate, who saw the Exarchate as an obstacle to the Hellenisation of the Orthodox population of
Macedonia and Thrace; the Serbs, who stood well with the Patriarchate,
and had similar aspirations to assimilate the Slavs of Macedonia; the
Roman Catholic Church, which had seized upon these intra-Orthodox
conflicts as an opportunity to proselytise among the Bulgarians, with
some local success; Austria-Hungary, the Catholics' patron, credited
with territorial ambitions in Macedonia; and more generally, the Western powers, who viewed the Bulgarian national movement as a tool of
Russian expansionism. The Exarch's own political strategy rested upon
three pillars: the willingness of the Ottoman government to accept that
continuing concessions to moderate, conservative Bulgarian nationalism were in its own political interest; the support of Russia, the Bulgarians' liberator, to whom they were tied by Orthodoxy and Slavdom, and

whose regional interests, the Exarch was confident, were fully compatible with Bulgarian national aspirations; and the continued revival of the Church, and the preservation of its moral authority over the Bulgarians.

II

As of 1878, however, the prospects for such a strategy were problematical. Ottoman confidence in the Bulgarians had collapsed, and with it, the Exarch's first pillar. The Sultan's government, blaming the Bulgarians for the war, and convinced that they represented a continuing threat to Ottoman security, moved swiftly to curtail the Exarchate's activities, ejecting its metropolitans from their Macedonian eparchies, harassing its churches, schools and community organisations, and toying, it was widely rumoured, with the notion of abolishing the Exarchate altogether. Not until late 1883 did the Sultan indicate that he would preserve the Exarchate; even so, the Macedonian eparchies were kept vacant, and the repression of the Exarchate's activities in the provinces continued, to the immediate advantage of the Oecumenical Orthodox Patriarchate and the Greeks. The Exarch, however, remained confident: patience, persistence and demonstrative loyalty would eventually wear down Ottoman resistance.[8]

Nor could the Exarch be entirely sure of his second pillar, Russia. As liberators, the Russians enjoyed great prestige and political influence in Bulgaria and Eastern Rumelia. For reasons that were both political and religious, however, they had long been unhappy with the schism between the Exarchate and the Patriarchate, and quickly suggested that the establishment of the two statelets might afford an opportunity for compromise: the Exarch should confine his jurisdiction to Bulgaria and Eastern Rumelia, withdraw from Constantinople to Sofia or Plovdiv, and resign the disputed regions of Macedonia and Thrace to the Patriarchate. The Exarch refused, insisting that compromise on these terms would mean abandoning the Bulgarians of the disputed regions to Hellenisation. For the time being, the Russians let the matter drop.[9]

It was the threat to the third pillar, the Church's own authority among the Bulgarians, which was to pose the most serious problems in the longer term. National political leadership had now passed to the governments of Bulgaria and Eastern Rumelia, and the Exarchate found itself substantially dependent upon them. Within their territories, which comprised the greater part of the Exarchate's jurisdiction, they had assumed its former responsibility for schooling and the upkeep of the clergy; they provided essential subsidies for the Exarchate's schools in Macedonia and southern Thrace; and the Exarchate's own relations with the Ottoman government were a hostage to theirs. Eastern Rumelia's governors were conservative and sympathetic to the Church. In Bulgaria, however, power had passed into the hands of parliamentary politicians, radicals and liberals, who did not share the Exarch's evolutionist views, and were, in many cases, irreligious and anti-clerical. Their tolerance of armed cross-border raids and anti-Ottoman agitation undermined the Exarch's efforts to regain the confidence of the Sultan's government, while their anti-clericalism was manifested in a series of disputes over the Church's funding, organisation and prerogatives. There were, too, disturbing signs of tension between these politicians and the Russians, who privately dismissed them as 'reds and atheists'.[10]

These politicians, the Exarch believed, were merely a symptom. They reflected the values and interests of the intelligentsia, Bulgaria's educated class, made up of some few thousand state officials and schoolmasters.[11] This intelligentsia was for the most part irreligious, its faith having been undermined by the protracted struggles against the Patriarchate and Catholic propaganda, and by foreign schooling and intellectual influences; it had no wish to accord the Church a prominent role in national life. Being generally without means, it looked to politics and the state for its livelihood. It was, in the Exarch's estimation, intellectually and morally depraved: driven by hunger and individual self-interest, lacking in fundamental loyalties, and prone to political adventurism. Its power derived from Bulgaria's democratic constitution, which enabled it to rule through demagogy, and also from the absence of countervailing forces: the Bulgarians had neither aristocracy, nor bourgeoisie, nor propertied middle class; the small conservative intelligentsia lacked influence; the clergy was poor and ill-educated; and the peasant mass was ignorant and easily misled. The Exarch's favoured solution was the adoption of a less democratic constitution, which would assure the ascendancy of conservative elements. In 1881 he expressed public approval when Prince Alexander, backed by Russia and the Sultan, suspended the constitution, though he was less approving of the Prince's decision to assume full powers.[12] In any case, within two years the Prince had quarrelled with the Russians and restored the constitution. The Exarchate was left isolated and vulnerable: without effective influence at Sofia, it could exert no leverage at Constantinople or St Petersburg.

III

From September 1885 onwards Bulgaria's unstable politics plunged the country into two years of domestic and external crises: commencing with the principality's absorption of Eastern Rumelia through a coup d'état, continuing with the violent overthrow of Prince Alexander and the severing of all relations with Russia, and ending, in August 1887, with the illegal installation of the Austrian interloper Ferdinand of Saxe-Coburg on the Bulgarian throne, without the requisite sanction of the European powers.[13] Initially, Exarch Yosif hoped that this series of upheavals might facilitate a reform of Bulgaria's constitution and the installation of a more conservative regime, perhaps assisted by a temporary Russian occupation of Bulgaria.[14] The eventual outcome was anything but satisfactory: a definitive political breach between Russia and Bulgaria, which turned to Austria-Hungary and Britain for support; the installation of a Catholic Prince in Bulgaria, and the establishment of a dictatorial regime supported by some '5–6,000 officials', reliant on 'thugs and the army', and led by the radical politician Stefan Stambolov, 'the personification', in the Exarch's words, 'of the Bulgarian extreme intelligentsia'.[15]

The Exarch knew he was powerless to oppose Stambolov's 'gang of anarchists'. Their power-base was the intelligentsia, and 'the intelligentsia is everything in Bulgaria': only an external force could overcome it.[16] The Russians, however, had abandoned Bulgaria in disgust, while the Turks were only too happy to see an anti-Russian regime at Sofia, whatever its domestic policies.[17] Worse, the Exarchate's own dependence upon the Sofia politicians had been increased by Bulgaria's absorption of Eastern Rumelia, and while the Exarch himself was shrewd enough to avoid giving direct offence to Stambolov and Ferdinand, some of his subordinate clergy were not. Stambolov soon showed his teeth, forcibly dispersing the principality's Synod, interrupting the flow of subventions for the Exarchate's educational work in Macedonia, and against all precedent, insisting upon inspecting the Exarchate's accounts.[18] It seemed to the Exarch that the forces of irreligion had triumphed. 'Atheism is settling in Bulgaria', he complained; 'Faith and the Church decline, the new generations emerging from the schools are nihilists.'[19] He even feared that the Stambolov regime, in its determination to shore itself up against Russia, might seek to turn Bulgaria Catholic.[20]

Such fears were exaggerated, for Stambolov could see the Exarchate's value as a vehicle for the Bulgarian cause in Macedonia. Over time, a *modus vivendi* was established, and in the event, it was thanks to the energetic support of Stambolov's government that the Exarch achieved his first real breakthrough in Macedonia: the Ottoman government's decision, in July 1890, to re-install Exarchist metropolitans at Skopje and Ohrid.[21] This was followed four years later by the Sultan's agreement to the installation of metropolitans at Veles and Nevrokop.[22] The Exarch noted optimistically that he had regained the ground lost since 1875, but he knew that in reality, the underlying Ottoman attitude had not changed, and that these concessions had been made for the sake of Stambolov's anti-Russian stance, not of a church whose dependence on Sofia had grown steadily more obvious over the years.[23] Since 1886, in fact, there had been mounting evidence that the Ottoman authorities were fostering Serb national propaganda in Macedonia, as a means of weakening Bulgarian sentiment, and dividing the Slav Orthodox population.[24] Disturbingly, there were also signs that some Russian officials in the region were encouraging Serb propaganda.[25]

IV

The Exarchate continued to live at the mercy of events, and the year 1894 saw the inception of a fresh period of political upheavals, commencing with the downfall – and subsequent murder – of Stefan Stambolov. This opened the way to a formal reconciliation between Bulgaria and Russia, concluded at the beginning of 1896, and to international recognition of Prince Ferdinand as Bulgaria's legitimate ruler.[26] The Exarch's response was cautious. For all his earlier hostility to Stambolov, he had come to acknowledge him as Bulgaria's first true statesman, who had successfully defended the country's independence against Russia, gained the support of the Western powers, and done some good to the Church. His dictatorial methods were regrettable, but their success confirmed the Exarch's opinion that Bulgaria was not ready for free institutions; the prospect of a return to unstable parliamentarianism did not appeal.[27] Nor was the reconciliation with Russia an unmixed blessing. For one thing, it was soon clear that Bulgaria had not returned to the Russian fold: rather, it seemed, the principality was left uncertainly poised between Russia and the Western powers, with no reliable source of international support.[28] For another, it also became clear that the Russians would not commit themselves to full support of Bulgarian national aspirations: they made a fresh attempt to broach the topic of healing the schism with the Patriarchate, and they gave increasingly explicit endorsement to Serb claims in Macedonia.[29]

1894 also saw the onset of a major Eastern Crisis, focusing initially upon the Armenian-inhabited regions of eastern Asia Minor. The prospect that the European powers might intervene to compel the Ottoman government to introduce a programme of administrative reforms in Asia Minor not unnaturally encouraged hopes that they might be induced to impose similar reforms on the Ottoman Empire's Balkan provinces; within Bulgaria, members of the influential Macedonian emigration organised a public agitation, under the direction of a Supreme Committee, for the establishment of an autonomous civil administration in Macedonia.[30] The Exarch responded sceptically. He saw no reason to believe that the Ottoman authorities would concede serious reforms, still less autonomy, of their own free will, or that the Great Powers would compel them to do so. Furthermore, he doubted that a purely civil or administrative reform would work to the advantage of the Bulgarians of Macedonia, who remained weak, backward in national consciousness, and vulnerable to Greek and Serb rivals. Better, in his opinion, to work for the consolidation of Bulgarian national consciousness through the Church and its schools, and to postpone any change in Macedonia's political status until the consolidation had been achieved; meanwhile it was pointless to irritate the Turks.[31] The government at Sofia, however, preferred to keep its options open. It encouraged the Exarch to press for fresh eparchies, but at the same time tolerated the domestic agitation for autonomy, and also turned a blind eye to the passage of armed bands across the Ottoman frontier. Meanwhile it engaged in secret negotiations with Serbia and Montenegro, with a view to joint diplomatic or military action, should circumstances warrant it. In the end, this flexible strategy paid off. By 1897 the Armenian issue was subsiding, and it was clear that the European powers would impose no significant reforms upon Macedonia. However, the outbreak of serious disturbances in Crete, leading to a brief war between Greece and the Ottoman Empire, offered Sofia the opportunity to sell its neutrality to the Sultan's government: its reward was the granting of three further Macedonian eparchies to the Exarchate, at Bitola (Monastir), Debar and Strumica.[32] The concession was balanced, however, by the Sultan's decision to accord generous school privileges in Macedonia to the Serbs.[33]

V

Bitola, Debar and Strumica proved to be the Exarchate's last gains. The Eastern Crisis subsided, but within Bulgaria, the agitation for Macedonian autonomy continued; meanwhile, unnoticed, some members of the Bulgarian intelligentsia within Macedonia set up an 'Internal Organisation', with the aim of achieving autonomy through armed insurrection.[34] In late 1897 the Ottoman authorities discovered caches of arms and explosives in eastern Macedonia; Exarchist priests and teachers were implicated, as, allegedly, was the Exarchist Metropolitan of Skopje.[35] The Exarch was much disturbed, for he was convinced that any insurrectionary movement would lead to disaster: either the Turks would smash it and, in the process, inflict incalculable damage on the Bulgarian community and cause in Macedonia, or else Austria-Hungary would seize the pretext to occupy Macedonia. Bulgaria, he foresaw, would be powerless to intervene without the support of a Great Power, and there was no reason to believe that such support would be forthcoming, particularly from Russia. The Exarch's own position was delicate enough: he could scarcely take sides against the Ottoman government, but nor could he openly condemn the revolutionaries, who enjoyed significant popular sympathy in Bulgaria and in Macedonia, and whose relationship to the Bulgarian government remained obscure. His instinct was to stand aside, and preserve the Church and as many as possible of the Macedonian Bulgarians from involvement. He urged the Sofia authorities to restrain the revolutionary movement, forbade his clergy and officials to assist it, and dismissed a number of school-teachers who were obviously compromised.[36]

The revolutionaries, however, would not allow the Church to stand aside. They needed to infiltrate its organisation, and particularly its schools, in order to obtain cover and influence among the population. Demands, backed by death threats, were made for the appointment of revolutionaries to posts in schools, and even for the resignation of senior Exarchist officials.[37] The Exarch was determined to resist this pressure, and did succeed, it seems, in preserving the central organs of the Exarchate from revolutionary influence. But he could offer little protection to clergymen, school directors and community heads in the Macedonian provinces: a few were murdered, some fled to Bulgaria, but most collaborated with the revolutionaries, out of sympathy or fear.[38]

The Exarch believed he understood the revolutionaries. They were young, educated persons, schoolmasters and the like, who saw no prospects for themselves in an Ottoman Macedonia, but believed that a political change would furnish them with careers. Their strategy, to provoke the Turks to massacres which would in turn provoke the powers to intervene, was adventuristic and naive: the Exarch foresaw that the Turks would not fall so easily into the trap, and no power would intervene to save the Bulgarians. Nor, in the Exarch's judgement, were the Bulgarians of Macedonia prepared for the ordeal facing them: national consciousness and its concomitant spirit of self-sacrifice remained weak, outside a few areas. The revolutionaries' ideals, too, were suspect. They were irreligious: teachers affiliated to the committees were promoting atheism and even socialism in the Exarchate's schools, and neglecting their duty to strengthen the Bulgarians through proper education. Further, the revolutionaries were unsound on national issues: some, at least, manifested a 'separatism' which appeared to regard the achievement of Macedonian autonomy as an end in itself, and the consolidation and unification of the Bulgarian nation as a secondary matter.[39] In sum, the Macedonian movement was a further manifestation of the vices of the Bulgarian intelligentsia; in Bulgaria, too, atheism, socialism and anti-national ideas were gaining ground, particularly among students and the teaching profession.[40]

Needless to say, the growth of the revolutionary organisation did nothing for the Exarch's relations with the Ottoman authorities. The Church's progress came to an abrupt halt, as the Sultan's ministers refused all requests for further eparchies or other concessions.[41] Worse, the authorities began to give fresh encouragment to the Serbs, tolerating their propaganda, permitting them to establish more schools, and even, in 1902, sanctioning the appointment of a Serb, Firmilian, as Patriarchist Metropolitan of Skopje.[42] All this was to some effect: in the same year, the Exarchate estimated that the Serbs had won over some 25 per cent of the Bulgarian population in the northern Macedonian districts of Kumanovo, Skopje, Tetovo and Gostivar, or 44 per cent, if the district of Presevo was included.[43] In vain did the Exarch warn the Ottoman government that these tactics were forcing the population into the arms of the committees, and that the safest way to neutralise the revolutionaries would be to make concessions to the Church and offer state employment to the intelligentsia: the Sultan's ministers retorted that the revolutionary organisations were manned by Exarchist teachers, priests and the graduates of Exarchist schools, and that by making concessions to the Exarchate in the past, they had created a rod for their own backs.[44]

Nor could the Exarch find protection at Sofia. He was convinced that it was within the power of the Bulgarian government to cut off the revolutionary agitation at source, by suppressing the committees and preventing the passage of men and arms across the frontier. Successive cabinets offered assurances, and even made occasional arrests, but it was manifest that this was only for show: the revolutionaries continued to agitate, and to expand their organisation in the Principality and in Macedonia.[45] All the Exarch's information suggested that it was Prince Ferdinand himself who was protecting the revolutionaries, and even financing them, not for the sake of liberating Macedonia, but as a means of shoring up his own domestic position. Ferdinand, the Exarch had decided, was an exceptionally dangerous man, 'more cunning than any Bulgarian', whose capriciousness, self-seeking and lack of fundamental loyalties were the mirror-image of the Bulgarian politicians and intelligentsia whom he had so successfully mastered.[46] Even the Russians could not be counted upon. They shared the Exarch's disapproval of the revolutionaries, and made their views known at Sofia; but they had also endorsed Firmilian's appointment at Skopje, and were hinting that Bulgarians and Serbs should reach an accommodation, on the basis of an agreed division of Macedonia into spheres of influence.[47]

Caught between the unyielding attitudes of the Ottoman and Bulgarian governments, and with Russia's attitude devastatingly plain, the Exarch could but await the inevitable explosion, which duly came with the Ilinden Rising in western Macedonia in August 1903.[48] The rising was a bloody failure; the Exarch feared that it had set back the Bulgarian cause in the affected regions by twenty years. Led by Russia and Austria-Hungary, the powers intervened to impose a programme of civil reforms in Macedonia, but made it plain that they would not raise issues of nationality. Meanwhile the Ottoman army, supported by armed Turkish, Greek and Serbian bands, embarked upon a campaign of attrition against the internal organisation and the Bulgarian population, harassing the clergy and intelligentsia, and forcibly promoting conversions to the Patriarchate. The Exarch repeatedly warned the Bulgarian government that it must curtail the revolutionary movement, or face the destruction of the Bulgarian population of Macedonia. His warnings were without effect: to Prince Ferdinand and his cabinets, it seemed, the Macedonian cause was too valuable a source of personal, political and even pecuniary advantage.[49] Not until 1906 did there come the first hints of a softening in the Ottoman government's attitude, by which stage the revolutionaries in Macedonia were on their last legs, and the local population was turning against them, incidentally freeing the Exarch to act against revolutionary influence in the schools.[50] But by then it was too late: violent death, flight, emigration and desertions to the Patriarchate had severely reduced the Bulgarian population of Macedonia, to a point where the Exarch feared that its status as the largest Christian community might be in jeopardy.[51]

VI

From 1906 onwards the Exarch's health began to fail, and he entered into a phase of semi-retirement. Not that he could afford to ignore public affairs entirely. 1908 saw a constitutional revolution in the Ottoman Empire, but the Exarch responded sceptically: he had never believed in the Turks' willingness to reform, and was not surprised when the new constitutional authorities revived their predecessors' tactic of promoting the Serb cause in Macedonia.[52] In some respects, the new regime was worse, for it commenced an assault on the privileges of the Christian churches, insisting that it would henceforth recognise them as spiritual institutions only. The upshot, in 1911, was a series of overtures from the Patriarchate, proposing the establishment of a common front, and hinting at a lifting of the schism. The Exarch responded positively, but warned that there could be no question of abandoning Bulgarian national aspirations in Macedonia.[53]

In reality, the Bulgarian cause in Macedonia was on the eve of catastrophe. In October 1912 Bulgaria joined forces with Serbia, Montenegro and Greece to launch the First Balkan War against the Ottoman Empire, easily driving the Turks from Macedonia and Thrace. Within months, however, the victors had quarrelled, and in June 1913 Bulgaria attacked Greece and Serbia, only to be attacked in turn by Romania and the Ottoman Empire. The result was a national disaster. Bulgaria lost the Dobrudzha to Romania, Eastern Thrace to the Turks, and nearly all of Macedonia to Greece and Serbia; her only gains were Western Thrace and a small corner of Macedonia. Russia offered no protection. The Exarch's presence in the Ottoman capital had become an anachronism, and in November 1913 he withdrew permanently to Sofia.[54]

The Exarch's verdict on this national catastrophe was implicit in the critique of Bulgarian society and politics which, over several decades, he had developed in his diaries. At bottom, he believed that the Bulgarian national revival had taken a wrong turn in 1878, when political and moral authority had passed from the Church to gangster politicians whose constituency was an irreligious and office-hungry intelligentsia. The intelligentsia's loss of faith had undermined moral restraints, promoted materialism and selfishness, reduced patriotism to a mask for personal and party interests, and delivered the Bulgarians into the hands of a man like Prince Ferdinand.[55] It had also weakened the link to Russia, without whose protection the Bulgarians were lost. The shortsighted and self-seeking policies pursued by successive Bulgarian governments had alienated the Russians and the Turks, created openings for the Greeks and the Serbs, crippled the Bulgarians of Macedonia, and led directly to the catastrophe of 1913. To this extent, the catastrophe was indeed a divine punishment for the Bulgarians' sins.[56] The problem had another side, however. Statehood had come upon the Bulgarians before they were prepared for it. The peasant mass remained politically and socially inert. There was no aristocracy, bourgeoisie or developed middle class, while the Church, bereft of political support, had been marginalised; the resulting vacuum of leadership had been filled by the intelligentsia. 'The people is a sheep', the Exarch had noted in 1888, and 25 years later he held that the Bulgarians were still not fit for full independence, suggesting to a Russian visitor that the better alternative would be a form of Russian protectorate, as in Finland.[57] He spent the final months of his life in vain efforts to promote a stable consensus among Bulgaria's politicians, and forestall Bulgaria's entry into the First World War on the side of the Central Powers. He died in 1915.[58]

Notes

1 Khristo Temelski (ed.), *Balgarski Ekzarkh Yosif I: Dnevnik* (Sofia, 1992). Hereafter cited as *Dnevnik*; all dates are Old Style.
2 Details of Exarch Yosif's early life and career in Mikhail Arnaudov, *Ekzarkh Yosif i Balgarskata kulturna borba sled s, zdavaneto na Ekzarkhiyata (1870–1915)*, I (Sofia, 1915), chs i–vi, passim.
3 The Exarchate's situation after the Russo-Ottoman war is comprehensively surveyed in Kiril Patriarch Balgarski, *Balgarskata Ekzarkhiya v Odrinsko i Makedoniya sled osvoboditelnata voyna 1877–1878. Tom parvi 1878–1885*, 2 vols (Sofia, 1969–70); cf., Arnaudov, *op. cit.*, chs vii–xii, passim.; M. Arnaudov, *Kam istoriyata na balgarskata Ekzarkhiya. I: dokumenti ot 1881 do 1890 godina* (Sofia, 1944).
4 *Dnevnik*, 16.10.78; 19.10.78; 11.11.78; 24.5.80; 10.1.81.
5 For the origins and early activities of the Exarchate see Zina Markova, *Balgarskata Ekzarkhiya 1870–1879* (Sofia, 1989); Thomas A. Meininger, *Ignatiev and the Establishment of the Bulgarian Exarchate 1864–1872: a Study in Personal Diplomacy* (Madison, 1970).
6 *Dnevnik*, 9.12.78.
7 This and the following paragraph are based on statements scattered throughout the Exarch's diaries. For representative examples of his views on the centrality of Orthodoxy and the need to build up the Church, see *Dnevnik*, 10.7.81; 21.9.86; 6.5.88. For his views on the Bulgarians' political immaturity, and the need to pursue an evolutionary policy, see e.g., *ibid.*, 11.11.78; 25 & 26.5.81; 1.7.81; 11.11.87; 13.2.02; 11.4.10. For his views on the threats posed by the Patriarchate, the Greeks and the Serbs, see e.g., *ibid.*, 11.11.78; 17.7.80; 4.4.85; 26.7.90; 16.7.92; 25.3.96. For his views on the Catholic threat and Austria-Hungary see e.g., *ibid.*, 3.2.80; 10.10.80; 11.4.89; 24.3.95; 18.10.95. Also see the following articles which the future Exarch published in 1871–2:

'Slovo za polzite ot chitalishtata', *Chitalishte*, 1(17) (1 June 1871); 'Svobodata i Dlazhnostta', *ibid.*, 1(18) (15 June 1871); 'Strastta', *ibid.*, 1(19) (1 July 1871); 'Sebelyubieto', *ibid.*, 1(20) (15 July 1871), and 1(23) (1 Sept. 1871); 'Internasional', *ibid.*, 2(3) (1 Nov. 1871); 'Belezhki ot edin znamenit anglichanin varkhu uchenieto i vaspitanieto', *ibid.*, 2(7) (30 Dec. 1871); 'Saglasovanie na vaspitanieto s razvitieto na dushevnite sposobnosti', *ibid.*, 2(9) (1 Feb. 1872); 'Oras Man', *ibid.*, 2(10) (15 Feb. 1872). Cf., Elena Traykova, 'Obshtestveno-politicheski vazgledi na Lazar Yovchev (Ekzarkh Yosif)', *Istoricheski pregled* (1983), no. 5, whose account of the Exarch's views differs radically from that presented here.

8   *Dnevnik*, 11.11.78; 24.11.78; 22.1.80; 1.4.81; 12.1.83; 27.1.83; 6.2.83; 2.10.83; 5.10.83; 17.12.83; 20.4.85.

9   *Ibid.*, 9.12.78; 22.10.80; 10.1.81; 17.1.81; 1.2.81; 20.3.81; 1.4.81; 15.4.81; 19.1.83; 2.12.83; 18.12.83; 17.3.84; 12.3.86; 16.7.86.

10   *Ibid.*, 9.10.78; 3.11.80; 24.1.81; 2.2.81; 19.3.81.

11   Comments on the Bulgarian intelligentsia are scattered throughout the Exarch's diaries. Among the more important are *Dnevnik*, 10.1.81; 21.9.86; 11.11.87; 6.5.88; 24.1.89; 22.9.90; 31.12.92; 6.12.98; 25.1.01; 1.10.07; 8.4.10; 11.4.10; 22.4.10; 29.3.15.

12   *Ibid.*, 21.5.81; 25–26.5.81; 1.7.81.

13   Background for the period 1885–94 in Simeon Radev, *Stroitelite na savremenna Balgariya*, 2 vols (Sofia, 1911); Dimitar Marinov, *Stefan Stambolov i noveyshata ni istoriya* (Sofia, 1909); Andrey Pantev, *Angliya sreshtu Rusiya na Balkanite 1878–1894* (Sofia, 1972); Joachim von Königslöw, *Ferdinand von Bulgarien. Vom Beginn der Thronkandidatur bis zur Anerkennung durch die Grossmächte 1886 bis 1896* (Munich, 1970); Arnaudov, *Kam istoriyata na balgarskata Ekzarkhiya. I.*

14   *Dnevnik*, 21.10.85; 5.11.85; 7.2.86; 12.2.86; 3.3.86; 21.3.86; 7.4.86; 2.9.86; 21.9.86; 2.10.86; 27.1.87.

15   *Ibid.*, 18.8.86; 10.6.87; 3.7.88.

16   *Ibid.*, 21.9.86; Elena Statelova and Radoslav Popov (eds), *Stefan Stambolov i negovoto vreme: nepublikuvani spomeni* (Sofia, 1993), p. 305.

17   *Dnevnik*, 3.1.88; 10.10.89.

18   *Ibid.*, 4.3.87; 7.3.87; 13.3.87; 23.7.87; 1.9.87; 3.10.87; 11.11.87; 1.11.88; 17.11.88; 6.1.89; 17.1.89; 24.1.89; 11.4.89.

19   Statelova and Popov, *op. cit.*, p. 331.

20   *Dnevnik*, 14.9.88; 9.6.90. This fear resurfaced in 1892–3, when Stambolov's government amended the constitution to permit a Catholic to succeed to the throne; *ibid.*, 31.12.92; 7.1.93.

21   *Ibid.*, 19.5.89; 15.10.89; 25.8.90; 22.9.90; 23.10.92; 'Der Konflikt des Okumenischen Patriarchats und des bulgarischen Exarchats mit der Pforte 1890', in Gunnar Hering, *Nostos: Gesammelte Schriften zur südosteuropäischen Geschichte* (Frankfurt am Main, 1995).

22   Bozhidar Samardzhiev, 'Politikata na osmanska Turtsiya kam knyazhestvo Balgariya (1888–1896)', *Studia Balcanica*, 16 (1982).

23   *Dnevnik*, 21.1.88; 12.12.91; 15.3.93; 27.2.93; 12.6.94.

24   *Ibid.*, 23.7.86; 2.2.91; 16.7.92; 5.10.92; 13.8.94; cf. Kliment Dzambazovski, *Kulturno-opstestvenite vrski na makedoncite so Srbija vo tekot na xix vek* (Skopje, 1960), ch. iii, *passim*.

25 *Dnevnik*, 30.4.87; 21.6.87; 26.6.87; 8.11.89; 3.2.90.
26 These upheavals are covered in Radoslav Popov, *Balkanskata politika na Balgariya, 1894–1898* (Sofia, 1984).
27 *Dnevnik*, 10.11.94; 7.7.95.
28 *Ibid.*, 10.11.94; 'Conclusion for this year 1895 ...'
29 *Ibid.*, 1.1.96; 25.3.96; 24.4.96; 25.4.96; 15.6.96; 25.6.96; 7.7.96; 12.7.96; 24.7.96; 29.8.96; 18.9.96; 5.10.96; 7.11.97.
30 Khristo Silyanov, *Osvoboditelnite borbi na Makedoniya*, I (Sofia, 1934), pp. 55ff.
31 *Dnevnik*, 2.3.95; 9.3.95; 24.3.95; 18.10.95; 15.2.96.
32 Popov, *op. cit.*, chs ii–iii *passim*; *Dnevnik*, 21.10.97.
33 Popov, *op. cit.*, p. 157; *Dnevnik*, 16.4.96; 30.10.96.
34 For the Macedonian struggle, see Khristo Silyanov, *Osvoboditelnite borbi na Makedoniya*, 2 vols (Sofia, 1933–43); Fikret Adanir, *Die Makedonische Frage: ihre Entstehung und Entwicklung bis 1908* (Wiesbaden, 1979); Milen Mikhov, 'Balgarskata Ekzarkhiya i Ilindensko-preobrazhenskoto vastanie 1903 g.', in L. Panayotov (ed.), *90 godini Ilindensko-preobrazhensko vastanie* (Sofia, 1994).
35 *Dnevnik*, 14.3.98.
36 *Dnevnik*, 9.1.98; 14.3.98; 17.3.98; 15.5.98; 28.4.99; 20.3.1900; 10.4.01.
37 *Ibid.*, 12.8.98; 17.8.98; 18.8.98; 20.8.98; 23.8.98; 29.12.99; 24.9.01.
38 *Ibid.*, 14.9.1900; 22.9.1900; 24.10.1900; 24.9.01; 31.3.03; 5.2.03.
39 *Ibid.*, 3.11.99; 1.6.1900; 14.9.1900; 22.9.1900; 24.10.1900; 18.11.1900; 23.1.01; 17.6.02; 4.2.03; 24.3.06; 6.7.06.
40 *Ibid.*, 25.1.01; 22.11.01.
41 *Ibid.*, 28.11.98; 20.2.99; 27.3.1900; 24.9.01; 25.1.02.
42 *Ibid.*, 18–20.3.98; 11.8.98; 8.5.99; 17.6.99; 23.10.99; 25.10.99; 27.10.99; 14.11.99; 29.12.99; 16.3.02; 16.6.02; 17.6.02.
43 *Ibid.*, 23.1.01; 29.1.01.
44 *Ibid.*, 25.5.01; 13.3.02; 2.3.03; 19.6.03; 26.9.03.
45 *Ibid.*, 31.12.98; 5.4.99; 29.4.99; 23.10.99; 25.10.99; 18.11.1900; 10.1.01; 12.4.01; 3.2.03; 16.5.03.
46 *Ibid.*, 31.12.98; 6.1.99; 11.8.99; 29.1.1900; 10.6.1900; 24.10; 1900; 16.11.01; 22.11.01; 11.1.02; 25.1.02.
47 *Ibid.*, 28.4.99; 26.10.99; 16.8.01; 18.3.02; 7.6.02.
48 *Ibid.*, 16.5.03; 22.7.03; 31.7.03.
49 *Ibid.*, 21.8.03; 22.12.03; 9.4.04; 5.7.04; 30.7.04; 16.9.04; 9.9.05; 27.10.05; 1.3.06; 14.3.06; 18.4.06; 4.8.06; 4.9.06; 12.3.07; 14.1.08.
50 *Ibid.*, 26.1.06; 4.5.06; 19.5.06; 26.7.06; 4.8.06; 18.9.06; 14.11.06; 4.4.07; 7.11.07.
51 *Ibid.*, 18.9.06; 14.11.06; 18.12.06; 3.4.07; cf. A[tanas] Sh[opov], 'Balgarskoto kulturno delo v Makedoniya v tsifri', *Balgarska zbirka*, 18(5) (1.5.1911).
52 *Dnevnik*, 7.3.09; 12.8.09; 7.9.09; 11.12.10; 15.12.10.
53 *Ibid.*, 1.1.10; 15.10.10; 8.12.10; 17.3.11; 27.4.11; 11.9.11; 5.5.12.
54 *Ibid.*, 20.9.13; 27.11.13. Bulgaria's participation in the Balkan wars is chronicled in Andrey Toshev, *Balkanskite voyni*, 2 vols (Sofia, 1919–31).
55 *Ibid.*, 11.4.10; 22.4.10.
56 *Ibid.*, 24.12.14.
57 *Ibid.*, 3.7.88; 6.8.13.
58 *Ibid.*, 26.5.14; 24.7.14; 27.7.14; 31.7.14; 3.11.14; 15.11.14; 25.3.15; 29.3.15.

13
Sharpened Minds: Religious and Mythological Factors in the Creation of the National Identities in Bosnia-Herzegovina

Mitja Velikonja

Pre-modern Bosnian religious history

The intention of this chapter is to explain how religious-national mythologies contributed to stirring up recent events in the central Balkans. The contemporary religious and national history of Bosnia-Herzegovina offers a perfect and tragic example of the extent and potency of the political abuse of religious-national myths.

Bosnia-Herzegovina is a country with an exceptionally rich religious history; religions have played an important role over the course of centuries. In the medieval Kingdom of Bosnia, which was particularly strong during the fourteenth century under the Kotromanić dynasty, three Christian denominations coexisted peacefully. Roman Catholics flourished mostly because of the presence of the Franciscan order. Orthodox Christians were concentrated in certain parts of the country (namely Hum, roughly the territory of present-day Herzegovina). The outstanding feature of the religious structure of that time was the existence of the autochthonous Bosnian Church with its own type of monastic organisation, independent of the Roman Catholic Church and papal control.

The Bosnian Church was therefore schismatic, but not heterodox: its teachings and doctrine remained Roman Catholic, although it was deliberately labelled dualist or Manichaean (Cathar or Patarin) by its opponents. Nineteenth-century Romantic historiography and literature mistakenly identified its members as Bogomils ('bogumili'), which was indeed a neo-Manichaean sect from the southern Balkans. In some periods high representatives of the Bosnian Church were closely connected with the Bosnian royal court, which struggled for Bosnian independence against the neighbouring Kingdom of Hungary. Hungarian military efforts were reinforced and justified by clear papal endeavours to subordinate the autonomous Church of this 'heretical country', as the popes of that time called Bosnia.

When the Ottomans conquered Bosnia-Herzegovina in the second half of the fifteenth century, the religious structure of Bosnia was dramatically altered. The Bosnian Church ceased to exist. Roman Catholic and Orthodox Christians were allowed to continue practising their religions. Several dozen families of Sephardic Jews emigrated to Bosnia from Spain and Portugal following the Christian 'reconquista'. But undoubtedly the most far-reaching religious process was the mass and unforced conversion of large parts of the Bosnian Christian population to Islam. Religious communities in the Ottoman Empire were not only protected but also strengthened under the *millet* system, which permitted their autonomy as well as the preservation and even the reinforcement of their religious, historical, cultural and ethnic heritage.[1]

Pre-modern Bosnian religious history could, therefore, be characterised as a history of religious diversity and divisions as well as a history of mutual religious tolerance and respect. There was no religious hatred, none of the large-scale confrontations, brutal conversions or religious wars typical of the history of Europe and the Near East at that time. Conflicts and violence between members of different religious denominations were more sporadic exceptions than the rule. Inter-religious tensions did not start until the late nineteenth century and were incited from outside the country: namely, by the Greater-Serbian and Greater-Croatian religious-national mythologies and nationalistic programmes. It was at this time – despite the universal character of these religions – that religious identity became the most significant basis for national homogenisation and identification.

The structure of mythology

Mythology is an organised set of beliefs, images, symbols and stereotypes held by a society about itself and others. In my opinion, mythology exists in both pre-modern and contemporary societies. There are two types of myth: first, 'traditional' myths which are part of the historical heritage of the society (subscribed to by all or most members of society) and oriented to the past (as the 'roots' of the society); and secondly, 'ideological' myths (or ideology) which embody the specific interests of some political or social groups and are oriented to the future.

Mythology unites these two types of myths – i.e. the 'spring' of the traditional myth and the 'derivation' of the ideological one – in a specific, dynamic, changing unity. The most important question that arises from an analysis of the mythology of a certain society concerns the relation of myth to historical truth and fact. Mythology is the construction as well as the perception of a specific reality within a society. Therefore, historic truth is less important. Indeed, in the majority of cases, historical truth is unimportant, ignored and even concealed.

There are three main functions of mythology in society: integrative (mythology is the integrative, mobilising force of a society, uniting people); cognitive (mythology explains all the most important aspects of social life and the meaning of the past, present and future of the society); and metaphorical or linguistic (some mythical expressions and syntagma often become part of the language and thus perpetuate certain mythic beliefs).[2]

Serbian and Croatian Christo-Slavic myths

The Serbian Orthodox Church took on the role of political representative of the predominantly rural Orthodox Serb population under Ottoman rule and from that time on has identified itself completely with Serbian national interests. It has a long tradition of the sacralisation of Serbian rulers, political leaders and war heroes. It is therefore understandable that it contributed decisively to the creation of contemporary Serbian national self-awareness. Serbia became a semi-independent principality in 1829 and continued to extend its territory up until the First World War. The new Serbian religious-national mythology was formed in a parallel process. It was composed of six principal religious-national traditional and ideological myths:

1   In keeping with the 'myth of Kosovo', modern Serbia was seen as the resurrection of the medieval Serbian state, lost after the decisive battle with the Ottomans on 28 (15) June 28 1389, which in reality ended in a draw[3] (see Mihaljčić, 1989; Vucinich and Emmert, 1991). Events, situations and personalities in the nineteenth and twentieth centuries were identified with their medieval antecedents (the figures of the ruler, Prince Lazar, the hero, Miloš Obilić, and the traitor, Vuk Branković) and 28 June, St Vitus Day ('Vidovdan'), was widely celebrated. Advocates

of Greater Serbia promoted the idea that almost all Balkan Slavs were in fact Serbs. We can find such viewpoints and Kosovo motifs in some works of the writer Dositej Obradović (1739–1811); the bishop, writer and historian Lukijan Mušicki (1777–1837); the writer and linguist Vuk Stefanović Karadžić (1787–1864); the writer Jovan Sterija Popović (1806–56); the historian Ljubomir Kovačević (1848–1918); the anthropologist and geographer Jovan Cvijić (1865–1927); and others. The Kosovo myth did not become the central Serbian religious-national myth until the early nineteenth century.[4] All wars fought by Serbs and Montenegrins from then on were seen as the chance finally to avenge the Kosovo 'defeat'.

2   The 'myth of the sleeping king, Kraljevich Marko', an historical character (in reality an Ottoman vassal) who was depicted as a redeemer, sleeping in the mountain (most frequently Šar mountain). He is waiting for the right time to awaken and defeat the invaders and, ultimately, to restore the glorious Serbian kingdom, including many neighbouring territories, above all certain parts of Bosnia-Herzegovina.

3   In accordance with 'Svetosavlje', the heritage of Saint Sava, founder of the autocephalous Serbian Orthodox Church in 1219, all Serbs are inevitably Orthodox Christians and thus permeated with the Orthodox tradition. On the basis of this mythical unity of the sacred and profane, the collective and the individual in the Serbian past and present, it is claimed that the Serbian Orthodox Church is 'the heart' of the Serbian national identity. 'Svetoslavlje' is regarded as 'Orthodoxy of Serbian style and taste'.[5] Far from being out of use in modern times, such assertions can be traced in the writings of some distinguished Serbian theologians and clerics of the twentieth century, such as Nikolaj Velimirović, Justin Popović, Irinej Bulović and Atanasije Jeftić.

4   Particularly prominent is the 'myth of betrayal': converts to Islam – i.e. southern Slavs of Muslim faith – were treated as traitors to their religion and to their nation who should be eliminated or, at best, banished or reconverted to 'the faith of their grandfathers' (for example, in the writings of Petar Petrović II Njegoš (1813–51), the Montenegrin religious and political leader ('vladika') and poet).

5   There exists also a fifth myth: that of the Serbs as a 'chosen' or 'heavenly' people, which was promulgated under Ottoman rule but endured into the nineteenth and twentieth centuries. Using the 'myth of victimisation', Serbs claim that, although innocent, they have suffered throughout their history.

6 Finally, in their religious-national mythology, Orthodox Serbs per-
ceive(d) themselves as the bulwark against the expansion of Islam from
the south-east and against the expansion of Roman Catholicism from
the north-west.

In the Croatian lands, there existed two main types of national
mythologies under Habsburg rule. Compared to the role of the Serbian
Orthodox Church in the Serbian nation-building process, the Roman
Catholic Church was by no means such an important and powerful
factor in the creation of modern Croatian national identity, although
it did acquire some religious dimensions.

The first type of Croatian national mythology, Illyrianism, was
founded on the old and false assumption that all southern Slavs (from
Slovenes to Bulgarians) were descendants of the ancient Illyrian tribe.
While some Illyrian ideas appeared from the sixteenth century onwards
(for example, expressed by the Dominican friars Vinko Pribojević (sixteenth century) and Juraj Križanić (1618–83), the Bishop of Zagreb,
Maksimilijan Vrhovec (1752–1827), the Franciscan Ivan Franjo Jukić
(1818–57)), it reached its height in the 1830s and 1840s under the
energetic political and cultural leadership of Ljudevit Gaj (1809–72)
and his followers (especially around the newspapers *Novine Horvatske*
and *Danicza*). This integrative notion of the unity of the southern
Slavs and the subsequent need for their co-operation or even fusion –
despite their religious differences – survived in 'Yugoslavism' which
emerged during the 1860s, particularly in the circle around the liberal,
enlightened and nationally minded Bishop of Djakovo Josip Juraj Strossmayer (1815–1905). In terms of religion, this type of national mythology was rather ecumenical and it called for the friendship, collaboration
and perhaps even unification of the Orthodox and Roman Catholic
Church. Strossmayer even met with the Orthodox Metropolitan of Belgrade, Mihajlo, as well as with the president of the Serbian government
and leading Greater-Serbian ideologue of that period, Ilija Garašanin
(1812–74). The religious dimension of Croatian Illyriansm and Yugoslavism was named the 'Cyril-Methodius' idea, after the ninth-century
missionaries of the southern and western Slavs, Cyril and Methodius,
symbols of their Christian solidarity.

The second type of Croatian national mythology appeared in the final
decades of the last century and had a clearer religious dimension,
although it had some significant predecessors (for example Pavel Ritter
Vitezović (1652–1713)). It was more nationally exclusive: the first point
of reference was the powerful Croatian state of the eleventh century

(under the mighty kings of the Trpimirović dynasty, Petar Krešimir IV and Dmitar Zvonimir), which lost its independence in 1102.

According to the 'myth of lost empire', all lands that once formed the Croatian Kingdom – including some parts of Bosnia – were regarded as for ever Croatian. A legalistic obsession with these ancient historical rights of the Croatian nation gave the name to this political option ('pravaštvo') and to many political parties (e.g. 'Hrvatska stranka prava'). The first leader of this political movement was Ante Starčević (1823–96), a disillusioned Illyrianist.

Thus, according to the myth-makers of the second Croatian religious-national myth, almost all southern Slavs, regardless of their present religious identity, were considered to be of Croat and Roman Catholic origins. It was said that Bosnian Slav Muslims were godless, being descendants of weak and degenerate Christians who had converted in a cowardly and opportunistic way to the religion of their conquerors, in order to protect their political position, social status and property. But at the same time, Croatian nationalists tried to invite and include them into the Croatian national body: Bosnian Slav Muslims were often courted as 'the purest Croats' or 'our brothers of Muslim faith'. In the mind of such thinkers, Bosnia Herzegovina was simply the 'heart of Croatia'.

The third mythical reference was complete and unconditional adherence to Roman Catholicism: myth-makers began to build up the crucial importance of the Roman Catholic faith in the construction of the Croatian national consciousness (for example, in slogans such as 'God and Croats').

Like several of their neighbours, from the times of the Ottoman attacks, the Croats saw themselves as 'antemurale Christianitatis', the last bastion of Christianity; and they also perceived themselves as the wall of Roman Catholic Europe against the 'barbarism' and 'Orientalism' of the Orthodox Christian world.

These old 'Christo-slavic' myths – if I may use Michael Sells's splendid neologism – were developed and reconstructed in a very specific way under the clear influence of the European Enlightenment, Herder's *Sturm und Drang* literary and ideological movement, and ideas of the French Revolution. They were ideologically renewed in the nineteenth century and preserved their integrative force and explanatory credit also in various periods of the twentieth century.

Meanwhile, the Bosnian Slav Muslims did not develop any national mythology or national consciousness, at least none comparable with that of the Serbs and Croats. I believe that the main reasons for this difference were: the supranational organisation of the Ottoman Empire; Islam itself, which has no rigid ecclesiastical structure, comparable to the organisation of the various Christian churches; the absence of a nationally minded intelligentsia, aristocracy and clergy; political pragmatism and opportunism; and, lastly, the fact that there was no external nation-building centre, which would initiate and promote national awareness among the Bosnian Slav Muslims. A minority of them even accepted the notion that they were simply a part of the Serbian or Croatian nation, but of a different faith from the majority of Serbs/Croats. Nevertheless, they were mostly well aware of their ethnic, historical and linguistic differences from the Ottomans and their resemblance to the neighbouring Slav Christians.

'Nationalisation' of religious identities

Bosnia-Herzegovina was occupied by Austro-Hungary in accordance with the Congress of Berlin in the summer of 1878 and annexed in October 1908. The new authorities – like all colonial empires – saw themselves as the ones who would civilise the chaotic country; they justified the occupation as their cultural and political mission and the restoration of peace, order and prosperity for the whole population of the country. In other words, they perceived their intervention in mythical categories (namely, binary oppositions) – to use Mircea Eliade's terms – as the passage from chaos to cosmos. Those forty years of Habsburg rule, up to the end of the First World War, coincided with the onset of processes of transformation of the inhabitants of Bosnia-Herzegovina from their earlier predominantly religious identities to more national identities.

Until the mid-nineteenth century, Bosnia-Herzegovina was largely nationally undifferentiated: the most apparent and exposed distinctions were economic and religious-cultural. The process of 'nationalisation' was mostly influenced by external factors (Greater-Serbian and Greater-Croatian politics and religious-national mythologies) and it was later reinforced by certain internal factors (political and economic organisations, cultural, artistic, educational and historiographical institutions etc.). But, in addition to nationalistically oriented schoolteachers, educated mainly in the Serbian and Croatian lands, there were nationally minded members of the Christian clergy who were among the most zealous and effective promoters of the growing national identification of Bosnian Orthodox and Roman Catholic Christians. On the one hand, both sides depicted the Ottoman period as essentially hateful, burdensome and devastating for their religious and national communities; while on the other, they both hoped and actively tried to persuade Slav Muslims of the region to join their national movements and communities.

Throughout the nineteenth century, Illyrianism was particularly strong among the Franciscans from Croatia and Bosnia-Herzegovina. The latter in particular were very familiar with the multi-ethnic and multi-religious environment of the country. Ivan Franjo Jukić even wrote a history of Bosnia under the pseudonym 'Slavoljub Bošnjak'; in Zagreb he edited the journal 'Bosnian Friend' (*Bosanski prijatelj*). The other prominent figures among the Franciscans were Grga Martić (1822–1905), a poet and educationalist, Paskal Buconjić and Didak Buntić. Further, Illyrianism promoted tolerance towards the Slav Muslims of Bosnia-Herzegovina[6] (Friedman, 1996: 40). In contrast to the Franciscan monks, the Roman Catholic secular clergy disseminated new ideological religious-national myths on the basis of traditional ones, with the keen support of the Bosnian clerical Archbishop of non-Bosnian origins Josip Stadler (1882–1918).

The rivalry between the Franciscan order (especially strong in Herzegovina: for example in the monasteries of Mostar and Široki Brijeg) and the secular clergy continued also in the field of politics (two political options and parties, clerical and liberal) and over the Roman Catholic ecclesiastical structure in Bosnia-Herzegovina. But, in the next decades, both options slowly absorbed Great-Croatian religious-national mythology and saw the country as essentially Croatian. Although the new authorities promised to respect the religious diversity of the population, there were many incidents involving the other two main religious groups concerning conversions to Roman Catholicism. Members of the Muslim and Orthodox communities began to fear and resist the insidious attempts of Catholic proselytising among the non-Catholic population of Bosnia-Herzegovina with public protests, petitions and demonstrations.

Through the radicalisation of the religious-national myths outlined above, Orthodox monasteries and theological seminaries (for example Prizren, Žitomislić and Sarajevo) became the focal points of a systematic raising of Serbian national consciousness. These institutions and agitators were supported by the politics of the expanding Serbian state, which mobilised all the power at its disposal, including the Orthodox Church, in order to achieve its territorial objectives. From a mythical perspective, the national struggle – first against the Muslim Ottoman state and then against the predominantly Roman Catholic Austro-Hungarian

Empire – became a 'sacred act', 'a holy mission', a consecrated battle for the faith, openly stimulated also by the Orthodox clergy. All the inhabitants of Bosnia-Herzegovina were seen as of Orthodox and Serbian origin, but as having been converted – forcibly or voluntarily, through opportunism – to the Muslim or Roman Catholic faith. In particular the Slav Muslims of the region were often regarded as 'the most important and the most able part of the Serbian nation' or as being 'of the purest Serbian blood'.

Realising the potential danger and the destructive power of religiously based national mythologies, the Austro-Hungarian authorities prohibited any kind of national expression or demonstrations until 1903. In two decades of centralised 'administrative absolutism', they tried to isolate the inhabitants of Bosnia-Herzegovina nationally, religiously and politically from their irredentist neighbours. In the years immediately following the occupation they strove to exercise control over all three major religious hierarchies, thus weakening their connections with external religious centres. The question of the Orthodox population was setled with the appointment of an Ecumenical Patriarch in March 1880: a new organisational structure of the Roman Catholic Church in Bosnia-Herzegovina was agreed in July 1881; 'Ulema Medzlis', the highest religious body for Muslims there, was established in October 1882.

Political and social groups, cultural and business institutions were not allowed to have national names. Instead, under the Austro-Hungarian administrator Benjamin von Kallay (1839–1903), between 1882 and 1903, the occupiers strove to promote a regional supra-religious consciousness among the population of Bosnia-Herzegovina, i.e. 'bošnjaštvo', 'Bosnianism'. This concept was intended to provide also a sense of loyalty among all three major religious-national groups to the dual monarchy. According to Kallay, all inhabitants of the country constituted one single nation, namely the Bosnian nation, irrespective of religious affiliation (Muslim, Orthodox, Roman Catholic or Jewish). They were all seen as Bosniaks, who spoke the Bosnian language; a common Bosnian history and tradition was systematically studied and promoted.

This concept attained most support from the Bosnian Muslims, especially in the circle around Mehmedbeg Kapetanović Ljubušak (1839–1902), the editor of the weekly *Bošnjak* (1891–1910). Some other important adherents of this option were Safvetbeg Bašagić, a poet; Hilmi Muhibić; and Rizabeg Kapetanović. Bosnianism was strongly attacked from the Serbian and Croatian sides as a purely ideological invention and an imperialistic instrument to facilitate the rule of the colonial administration. Kallay's attempt failed and his successors had to accept the growing national and political differentiation of the population.

Bosnian religious pluralism led to political pluralism (the creation of Muslim, Serbian and Croatian political parties in Bosnia-Herzegovina), but it did not lay the foundation for democratic development and mutual tolerance among these different, many-sided groups. Instead, it became the source of national identification which was directed outside Bosnia-Herzegovina and, at the same time, against other nations within it. By the end of the century, Orthodox Christians ('Vlasi', 'hriščani' or 'hristjani') began to identify themselves as Serbs; Roman Catholics (Latinci, Šokci, krstjani or krščani) as Croats.

Only the Slav Muslims 'Poturci', 'Poturi' or 'Poturčenjaci' insisted on the primacy of religious over national identity. As distinct from Serbian or Croatian political programmes, those of the Muslim societies, movements and parties in the Austro-Hungarian period promoted only their autonomous economic, religious and cultural interests and on no account national ones. Serbian and Croatian nationalists were well aware that they could achieve dominance not only in Bosnia-Herzegovina but also in the Kingdom of Yugoslavia (1918–41) and in socialist Yugoslavia (1945–91) only if they incorporated the Slav Muslims of Bosnia-Herzegovina into their nations (as their 'compatriots of Muslim faith'). Muslims were not finally recognised as an independent national group until 1968 by the socialist Yugoslav authorities, which gave them, paradoxically, a religious name: Muslims ('Muslimani').

Survival of a Bosnian religious mosaic?

These newly imported exclusivist religious-national mythologies sowed discord and intolerance and usurped the traditionally multi-religious and synchretistic atmosphere of Bosnia-Herzegovina. Most of the Bosnian Orthodox and Roman Catholic Christians began to identify with the 'parent nations' outside Bosnia-Herzegovina, the Serbs and the Croats, respectively. The unavoidable consequences of this religious-national homogenisation and exclusivism were growing discomfort and distrust towards other religious and national communities within the country. From this perspective, former compatriots of different religious and national affiliation became 'eternal enemies' and a constant threat which had to be exterminated once and for all. In the militant perspective of the new religious-national ideological myths, differences between the inhabitants of Bosnia-Herzegovina became the source of all troubles. The final outcome was increased religious and consequently national hatred and tensions within Bosnia-Herzegovina, characterised by myth-makers as 'age-old', which contributed to the calamitous clashes of the Second World War and the 1992–5 war.

Religious-national mythologies were of course not the main reason for the outbreak of the most recent war. The major causes lay elsewhere: Greater-Serbian expansionism and territorial aggression, with the clear goal of uniting all Serbs in one state, no matter what the cost; and, to a lesser degree, the Croatian appetite for certain parts of Bosnia-Herzegovina. These myths helped to radicalise the identification of national and religious affiliation inwards and their incitement outwards, against others. In this way, they became a useful backdrop for the invading political and military efforts.

Notes

1   Zachary T. Irwin, 'The Fate of Islam in the Balkans: a Comparison of Four State Policies', in Pedro Ramet (ed.), *Religion and Nationalism in Soviet and East European Politics* (Durham and London: Duke University Press), pp. 378–40, p. 380; Wayne S. Vucinich, 'The Nature of Balkan Society Under Ottoman Rule', *Slavic Review*, 1962, no. 4, pp. 597–616 and pp. 633–8.

2   Mitja Velikonja, *Masade duha. Razpotja sodobnih mitologij* (Ljubljana: Znanstveno in publicistično središče, 1996), pp. 13–30.

3   Dragoljub Djordjević, 'The Serbian Orthodox Church: the Disintegration of the Second Yugoslavia, and the War in Bosnia Herzegovina', in Paul Mojzes (ed.), *Religion and the War in Bosnia* (Atlanta, Georgia: American Academy of Religions, Scholars Press, 1998), pp. 150–9, p. 153; Geert Van Dartel, 'The Nations and Churches in Yugoslavia', *Religion, State and Society*, no. 3, 4, 1992, pp. 275–88, pp. 281, 282.

4   Michael Sells, *The Bridge Betrayed. Religion and Genocide in Bosnia* (University of California Press, 1996).

5   Francine Friedman, *The Bosnian Muslims: Denial of a Nation* (Boulder: Westview Press, 1996), p. 40.

6   Enver Redžić, *Društveno-istorijski aspekt 'nacionalnog opredjeljivanja' Muslimana Bosne i Hercegovine* (Belgrade: Socializam, 1961), pp. 48, 49; Robert J. Donia, *Islam Under The Double Eagle: the Muslims of Bosnia and Herzegovina, 1878–1914* (Boulder, New York: East European Monographs, 1981), p. 15.

References

Alexander, Stella, *Religion and National Identity in Yugoslavia*, Occasional Papers on Religion in Eastern Europe (OPREE), 1, 1983, pp. 1–19.

Banac, Ivo, *Nacionalno pitanje u Jugoslaviji. Porijeklo, povijest, politika* (Zagreb: Globus, 1988).

Braude, Benjamin and Lewis, Bernard (eds), *Christians and Jews in the Ottoman Empire: the Functioning of a Plural Society*, Vol. I (London, Holmes and Meier, 1988).

Dartel, Geert van (1992), 'The Nations and Churches in Yugoslavia', *Religion, State and Society*, 3(4), 1992, pp. 275–88.

Donia, Robert J., *Islam Under Double Eagle: the Muslims of Bosnia and Herzegovina, 1878–1914*, East European Monographs (Boulder, New York, 1981).

Djordjević, Dragoljub, 'The Serbian Orthodox Church, the Disintegration of the Second Yugoslavia, and the War in Bosnia Herzegovina', in Paul Mojzes (ed.), *Religion and the War in Bosnia*, American Academy of Religions (Atlanta, Georgia: Scholars Press, 1998), pp. 150–9.

Dyker, David A., 'The Ethnic Muslims of Bosnia – Some Basic Socio-Economic Data', *Slavonic and East European Review*, 19, London, 1972, pp. 238–56.

Fine, John V. A., *The Bosnian Church: a New Interpretation*, East European Monographs, 10 (Boulder, New York and London, 1975).

Friedman, Francine, *The Bosnian Muslims: Denial of a Nation* (Boulder: Westview Press, 1996).

Hadžijahić, Muhamed, *Islam i muslimani u Bosni i Hercegovini* (Sarajevo: Svjetlost, 1977).

Handžić, Adem, *Population of Bosnia in the Ottoman Period: an Historical Overview* (Istanbul: IRCICA, 1994).

Hosking, Geoffrey and Schöpflin, George (eds), *Myths and Nationhood* (London: Hurst and Company; SSEES, University of London, 1997).

Irwin, Zachary T., 'The Fate of Islam in the Balkans: a Comparison of Four State Policies', in Pedro Ramet (ed.), *Religion and Nationalism in Soviet and East European Politics* (Durham and London: Duke University Press, 1989), pp. 378–407.

Imamović, Mustafa, 'Integracijske ideologije i Bosna', *Erasmus*, 18, Zagreb, 1996, pp. 38–47.

Kapidžić, Hamdija, *Bosna i Hercegovina u vrijeme austrougarske vladavine* (Sarajevo: Svjetlost, 1968).

Lopasic, Alexander, 'Bosnian Muslims: a Search for Identity', *Brismes Bulletin*, 2, 1981, pp. 115–25.

Lukić, Reneo, 'Greater Serbia: a New Reality in the Balkans', *Nationalities Papers*, 1, 1994, pp. 49–70.

Malcolm, Noel, *Bosnia: a Short History* (London: Papermac, 1996).

Mihaljčić, Rade, *The Battle of Kosovo in History and Popular Tradition* (Belgrade: BIGZ, 1989).

Miller, Nicholas, 'Two Strategies in Serbian Politics in Croatia and Hungary Before the First World War', *Nationalities Papers*, 2, 1995, pp. 327–51.

Mojzes, Paul, *Yugoslavian Inferno (Ethnoreligious Warfare in the Balkans)* (New York: Continuum, 1994).

Mojzes, Paul (ed.), *Religion and the War in Bosnia* (Atlanta: Scholars Press, 1998).

Norris, Harry Thirlwall, *Islam in the Balkans: Religion and Society Between Europe and the Arab World* (London: Hurst and Company, 1993).

Pinson, Mark (ed.), *The Muslims of Bosnia-Herzegovina*, Harvard Middle Eastern Monographs XXVIII (Cambridge: Harvard University Press, 1996).

Popović, Alexandre, 'La communauté Musulmane de Yougoslavie sous le régime communiste: Coup d'oeil sur son histoire et sur les principales institutions', in Patrick Michel (ed.), *Les réligions à l'est* (Paris: Cerf, 1992), pp. 174–81.

Purivatra, Arif, *Nacionalni i politički razvitak Muslimana* (Sarajevo: Svjetlost, 1969).

Ramet, Pedro, 'Primordial Ethnicity of Modern Nationalism: the Case of Yugoslavia's Muslims', *Nationalities Papers*, 2, 1985, pp. 165–87.

Ramet, Pedro, 'The Serbian Orthodox Church', in Pedro Ramet (ed.), *Eastern Christianity and Politics in the Twentieth Century* (Durham and London: Duke University Press, 1988), pp. 232–48.

Ramet, Sabrina Petra, 'Nationalism and the "Idiocy" of the Countryside: the Case of Serbia', *Ethnic and Racial Studies*, January 1996, pp. 70–87.

Redžić, Enver, 'Društveno-istorijski aspekt "nacionalnog opredjeljivanja" muslimana Bosne i Hercegovine', *Socijalizam*, 3, Beograd, 1961, pp. 31–89.

Schöpflin, George, 'The Ideology of Croatian Nationalism', *Survey*, 86, 1973, pp. 123–46.

Sells, Michael A., *The Bridge Betrayed: Religion and Genocide in Bosnia* (California: University of California Press, 1996).

Šidak, Jaroslav, *Studije o 'Crkvi Bosanskoj' i bogomilstvu* (Zagreb: Liber, 1975).

Velikonja, Mitja, *Masade duha. Razpotja sodobnih mitologij* (Ljubljana: Znanstveno in publicistično središče, 1996).

Velikonja, Mitja, 'Liberation Mythology: the Role of Mythology in Fanning War in the Balkans', in Paul Mojzes (ed.), *Religion and the War in Bosnia* (Atlanta: Scholars Press, 1998), pp. 20–42.

Velikonja, Mitja, *Bosanski religijski mozaiki – Religije in nacionalne mitologije v zgodovini Bosne in Hercegovine* (Ljubljana: Znanstveno in publicistično središče, 1998).

Voje, Ignacij, *Nemirni Balkan. Zgodovinski pregled od 6. do 18. stoletja* (Ljubljana: DZS, 1994).

Vucinich, Wayne S., 'The Nature of Balkan Society Under Ottoman Rule', *Slavic Review*, 4, 1962, pp. 597–616 and pp. 633–8.

Vucinich, Wayne S. and Emmert, Thomas A. (eds), *Kosovo – Legacy of a Medieval Battle*, Minnesota Mediterranean and East European Monographs (Minneapolis: University of Minnesota, 1991).

Appendix

We reproduce here an article which appeared in *St Mark, Monthly Review*, published by the Monastery of St Macarius the Great, Wadi El-Natrun, Egypt, May 1999, pp. 37–9.

Joint meeting held between representatives of the various religious denominations in Kosova, for the sake of peace in the region[1]

The International Association of Religions for the Cause of Peace arranged a meeting between the various religious denominations in Kosova in order to discuss ways of establishing peace in the region. That occurred last March (1999), in the capital, Prishtinë (and it pre-dated the air attacks which are now being launched by NATO against Yugoslavia). The participants at this meeting represented the Serbian Orthodox Church, the Roman Catholic Church and the Muslim community in the south-west of Serbia.

The participants confirmed in their final statement, issued after this meeting, that: 'All peoples have the right to live and earn a livelihood in Kosova and this region should never be exclusive to one community, yet denied to another.' Likewise, the participants launched an appeal for a peaceful solution to the crisis, calling on all to follow the path of negotiation and deploring all acts and attempts which were aimed at changing the course of this crisis, turning it from a political crisis into a religious war.

At this meeting, the Serbian Orthodox Church was represented by Father Sava, the head of one of the monasteries located in north-eastern Kosova. He read to those who attended a detailed statement prepared by Archbishop Artemije, the head of the Serbian Orthodox Church in Kosova. In this statement, he presented a clear idea and comprehensive picture of the possibility of allowing peace and tolerance to exist in Kosova. He also confirmed that the basic aim of convening this meeting was to investigate the human and secular problems to which the province of Kosova was exposed and which it was experiencing at this time. The aim of the meeting was not in any way to investigate the theological differences between the various denominations, nor to spur any ecumenical initiative of any kind.

In this statement of Archbishop Artemije's, it was affirmed that the Serbian Orthodox Church demanded a speedy cessation of violent acts and that the crisis should be resolved by peaceful means through a joint dialogue between the disputing parties, without the imposition of any prior conditions. The Archbishop launched a special appeal on behalf of the thousands of refugees, abducted individuals and those who had been arrested, likewise on behalf of all those who had been forced to leave their homes or to abandon them and who had lost their possessions and their personal property, confirming, with emphasis, that it was necessary to show respect for the basic and simplest rules in regard to human rights.

Continuing his statement, the Archbishop of Kosova said that: 'We condemn the desecration, damage and grave losses which have befallen religious edifices, be they churches, mosques or cemeteries. And we demand an obligation to protect especially houses of worship and archaeological sites which represent a cultural and historical value, and which are important for the country. That is because the ruination of this cultural and historical heritage – which has made this part of the Balkans a unique and renowned region at a global level – is something that cannot be tolerated under any circumstances. So too, the religious establishments must always continue to be a sanctuary and haven for tolerance and for peace, and any forms of destruction aimed against them should cease forthwith.'

The Serbian Archbishop continued his discourse by announcing his total rejection of the use of buildings and religious structures for other aims, without link or tie with either religion or humanity. He expressed his very strong determination and his staunch opposition to all kinds of religious discrimination and ethnic distinction, likewise to all acts of suppression, oppression and violation of the laws of human rights.

Archbishop Artemije repeated his statement that it was of prime importance to seek an early diplomatic solution to the problem of Kosova through peaceful negotiations between official representatives of the various peoples who inhabit the soil of the province, on condition that the discussions take place between the different parties within a framework of mutual respect and in a context framed by the rules of democracy and the recognition of the rights of man. In this situation, the Archbishop expressed his personal willingness to come forward and to participate personally in this peace effort. His Holiness is of the opinion that despite the presence of many differences, in language, history, culture and belief between the disparate peoples who reside in Kosova, it is in the common interest of all the communities that they should live together in peace and love, in order that it might be possible to guarantee a peaceful life for every individual and security and protection of his private and personal possessions.

The final statement indicated that the Prishtinë meeting had taken place in an atmosphere of mutual respect and in a spirit of tolerance. Its debates had been characterised by 'frankness, integrity and a worthy and well-intentioned view of differing shades of opinion', despite the fact that the discussions were not devoid of at times tough and serious debate. Even so, the participants expressed their pleasure that such a joint meeting as this had been convened amongst themselves, with the aim of discussing matters and investigating concerns which occupied all their minds in such a direct way.

The representatives of the three chief religious denominations in Kosova acknowledged a consensus of opinion that the situation in the region was 'abnormal' and that it was imperative to do 'something' as soon as possible to curb and put an end to the acts of violence, describing these as the primary cause of the great concerns to which all the people living on Kosova's soil were exposed. Every community was firm that its religious principles and observances confirmed that the human race is the 'most precious deputy of God Himself' (*athman khalifat Allah*). From this starting point all were certain of the need to stand together as one man and that should equally strive and endeavour to safeguard the rights of all the people regardless of their ethnic origins or their religion.

The participants at the Prishtinë meeting announced their total rejection of any attempt to exploit religion and to use it for political aims, to serve the interests of any party. Furthermore, they implored every party to the dispute in the region to stop using religious symbols, or hiding behind them, in order to further deeds of violence and fanaticism between people.

In conclusion, the Orthodox, Catholic and Muslim representatives welcomed a joint idea to hold further meetings of this kind, adopting this meeting as a starting point for their expression of a shared wish to create bridges of love in order to establish direct connections between each other. And to bring about a lowering of the intensity of tension in the region, trusting that this means would facilitate a limitation of the damage and avoid losses which had multiplied since the people of this region had begun to live in conflict and confrontation one against the other on Kosovan soil.

Note

1 This article (translated from the Arabic by Harry Norris) clearly reflects a view of the conference as seen from the viewpoint of the Serbian Orthodox Church. However, this journal expresses the view of the Ecumenical wing of the Coptic Church. It will be observed, though, that Kosova is spelt in the Albanian manner throughout (Kusufa). This, together with certain 'Islamic' turns of phrase, no doubt reflects the influence of the Egyptian press and media. Even so, Kusufu and Kusufa are both found in the Arab press in the Middle East and both forms appear in Dr Muhammad Mufaku's latest book, *Kosovo-Kosova: the Central Focus of the Albanian and Serbian Conflict in the Twentieth Century (Kusufu-Kusufa, bu'rat al-niza 'al-Albani-al-Sarbi fi'l-qarn al-ishrin)*, The Civilization Center for Political Studies, Markaz al-hadara lil-dirasat al-siyasiyya (Cairo, 1998).

Select Bibliography

Christianity in the Balkans

Alexander, Stella, *Church and State in Yugoslavia since 1945*, Cambridge: Cambridge University Press, 1979.

Chadwick, Henry, *The Early Church*, Pelican History of the Church, vol. 1, Harmondsworth: Penguin Books, 1967.

Dvornik, Francis, *Byzantine Missions among the Slavs*, Rutgers University Press, 1970.

Fine, John, V. A. Jr., *The Bosnian Church: a New Intrpretation*, East European Monographs, Boulder and New York, 1975.

Hamilton, J. and B. Hamilton, *Christian Dualist Heresies in the Byzantine World, c.650–c.1405*, Manchester: Manchester University Press, 1998.

Heppell, Muriel, *The Ecclesiastical Career of Gregory Camblak*, London, 1979.

The Jesus Prayer, by a Monk of the Eastern Church, Revised translation, with a Foreword by Kallistos Ware, New York: St Vladimir's Seminary Press, 1987.

Meyendorff, John, *Gregory Palamas and Orthodox Spirituality*, New York: St Vladimir's Seminary Press, 1974.

——*Byzantine Theology*, Revised edition, New York, 1983.

Obolensky, Dimitri, *The Bogomils: a Study in Balkan Neo-Manichean Dualism*, Cambridge: Cambridge University Press, 1948.

——*The Byzantine Commonwealth*, London, 1971.

The Life of Paisij Velychkovs'kyi, translated by J. M. E. Featherstone, with an introduction by A.-E. N. Tachaios, Cambridge, Mass: Harvard University Press, 1989.

The Philocalia (An English translation of the complete Greek text is now in progress, by G. E. H. Palmer, the late P. E. Sherrard and Kallistos Ware. Four volumes have already been published, vol. V is in progress.)

Runciman, Steven, *The Great Church in Captivity*, Cambridge: Cambridge University Press, 1968.

Sherrard, Philip, *Church, Papacy and Schism: a Theological Enquiry*, London: SPCK, 1978.

Southern, R. W., *Western Society and the Church in the Middle Ages*, Pelican History of the Church, vol. 2, Harmondsworth: Penguin Books, 1970.

Spinka, Matthew, *A History of Christianity in the Balkans* (first published Chicago, 1933, reprinted 1968; old, but still useful as an introductory work).

Vidler, Alec R., *The Church in an Age of Revolution*, Pelican History of the Church, vol. 5, Harmondsworth: Penguin Books, 1972.

Vlasto, A. P., *The Entry of the Slavs into Christendom*, Cambridge: Cambridge University Press, 1970.

Ware, Timothy, *The Orthodox Church*, revised edition, with an extensive bibliography, Harmondsworth: Penguin Books, 1997.

Wybrew, Hugh, *The Orthodox Liturgy*, London: SPCK, 1989.

Islam in the Balkans

Balic, Smail, 'Cultural Achievements of Bosnia-Hercegovinian Muslims', in F. H. Eterovich (ed.), *Croatia, Land and People*, vol. 11, University of Toronto Press, 1970, pp. 299–361.

Clayer, Nathalie, *L'Albanie, pays des derviches. Les ordres mystiques musulmans en Albanie à lépoque post-ottomane (1912–1967)*, Osteuropa-Institut der Freien Universität Berlin Balkonologische Veröffentlichungen, Berlin, 1990, vol. 17.

Clayer, Nathalie and Popovic, Alexandre, 'A New Era for Sufi Trends in the Balkans', *ISIM Newsletter* (International Institute for the Study of Islam in the Modern World), Leiden, no. 3, 1999, p. 32.

Daniel, Odile, 'The Historical Role of the Muslim Community in Albania', *Central Asian Survey*, London, 1990, vol. 9, no. 3, pp. 1–28.

De Jong, 'The Muslim Minorities in the Balkans on the Eve of the Collapse of Communism', *Islamic Studies*, Islamic Research Institute, Islamabad, 1997, vol. 36, nos. 2/3.

Duijzings, Ger, 'The End of a "Mixed" Pilgrimage', *ISIM Newsletter* (International Institute for the Study of Islam in the Modern World, Leiden, no. 3, 1999.

—— *Religion and the Politics of Identity in Kosovo*, London: C. Hurst, 2000.

Hasluck, *Christianity and Islam under the Sultans*, Oxford 1929.

Izetbegović, Alija, *Islam Between East and West*, Plainfield, Indiana: American Trust Publishers, 1984.

Kiel, Machiel, *Studies in the Ottoman Architecture of the Balkans*, London: Variorum Reprint, 1990.

Lockwood, W. G., *European Moslems: Economy and Ethnicity in Western Bosnia*, New York and San Francisco: Academic Press, 1976.

Matkovski, Alexander, 'L'Islam aux yeux des non-Musulmans des Balkans', *Balcanica*, Belgrade, 4, 1973, pp. 203–11.

Mélikoff, Irène, *Hadji Bektach, un myth et se avatars. Genèse et évolution du soufisme populaire en Turquie*, Leiden: Brill, Boston: Koln, 1998.

Norris, H. T., *Islam in the Balkans. Religion and Society between Europe and the Arab World*, London: C. Hurst, 1993.

Pašić, Amir, *Islamic Architecture in Bosnia and Hercegovina*, Studies in the History and Culture of Bosnia and Hercegovina, no. 2, Organisation of the Islamic Conference, trans. Midhat Ridjanović, Istanbul: Research Centre for Islamic History, Art and Culture, 1994.

Pinson, Mark (ed.), *The Muslims of Bosnia and Hercegovina. Their Historic Development from the Middle Ages to the Dissolution of Yugoslavia*, Center for Middle Eastern Studies, Harvard Middle Eastern Monographs, 28, Cambridge: Harvard University Press, 1993.

Popović, Alexandre, *L'Islam Balkanique. Les Musulmans sud-est européens dans la période post-Ottomane*, Osteuropa-Institut der Freien Universität Berlin Balkonologische Veröffentlichungen, Berlin, 1986, vol. 11.

—— *Un Ordre de Derviches en Terre d'Europe: La Rifaiyya*, L'Age d'Homme, Lausanne, 1993.

Sells, Michael, *The Bridge Betrayed: Religion and Genocide in Bosnia*, California: University of California Press, 1998.

—— 'Balkan Islam and the Mythology of Kosovo', *ISIM Newsletter* (International Institute for the Study of Islam in the Modern World, Leiden, no. 3, 1999), p. 31.

Shah-Kazemi, Reza (ed.), *Bosnia: Destruction of a Nation, Inversion of a Principle,* Islamic World Report, London, 1996.

Zulifkarpašić, Adil (in dialogue with Milovan Djilas and Nadezda Gace; with an introduction by Ivo Banac), *The Bosniak,* London: C. Hurst, 1998.

——'Islam in the Balkans', *Journal of Islamic Studies,* Oxford: Oxford University Press, vol. 5, no. 2, 1994.

Index

Printed in the United States
By Bookmasters